Time-encoding VCO-ADCs for Integrated Systems-on-Chip

Georges Gielen
Luis Hernandez-Corporales
Pieter Rombouts

Time-encoding VCO-ADCs for Integrated Systems-on-Chip

Principles, Architectures and Circuits

Springer

Georges Gielen
KU Leuven
Leuven, Belgium

Luis Hernandez-Corporales
Carlos III University of Madrid
Madrid, Spain

Pieter Rombouts
Ghent University
Ghent, Belgium

ISBN 978-3-030-88069-9 ISBN 978-3-030-88067-5 (eBook)
https://doi.org/10.1007/978-3-030-88067-5

This Springer imprint is published by the registered company Springer Nature Switzerland AG
The registered company address is: Gewerbestrasse 11, 6330 Cham, Switzerland

Contents

Part I

VCO-ADC Basics

Chapter 1
Time Based and VCO-ADCs from a Signal Processing Perspective

G. Gielen et al., *Time-encoding VCO-ADCs for Integrated Systems-on-Chip*,
https://doi.org/10.1007/978-3-030-88067-5_1

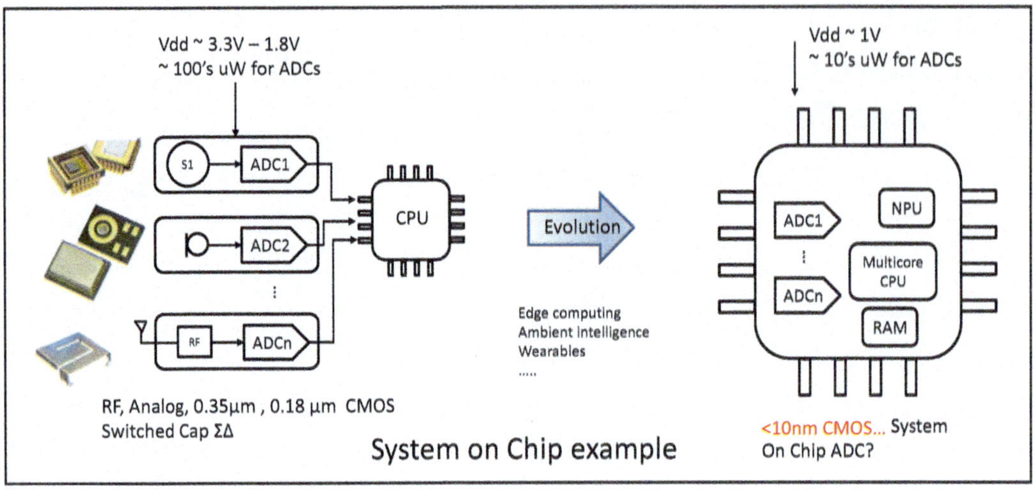

Fig. 1.1 Why use time encoding?

Nowadays consumer electronics is mostly driven by low-power ubiquitous applications like mobile communications, Internet of Things (IoT), or wearable devices. Most of these applications require simultaneously high-precision analog circuitry to interface sensors, complex digital circuits for data processing, and RF circuitry for communications. One of the problems that industry faces today is that each of these circuit flavors has its own preferred CMOS technology. Physical MEMS sensors like accelerometers, gyroscopes, temperature and pressure sensors, or microphones are bonded to signal conditioning and data conversion of CMOS chips in submicron technologies such as 0.18 μm or 0.35 μm and packaged as stand-alone modules. RF circuits are progressively being ported from BiCMOS into pure CMOS. Digital circuits are implemented in high-density CMOS technologies below 16 nm. To be power and cost effective, a natural evolution of this scenario is a progressive integration of high-density digital logic with more and more analog functions in a System on Chip (SoC) (Fig. 1.1).

The shrinking supply voltage, and the high mismatch and noise in nanometer CMOS technologies limit the dynamic range, causing conventional analog circuits to be hardly feasible in these technologies. Analog circuits based on time encoding [Roz97], [Laz04], and the concept of analog/digital signal processing [Tsi06] have been developed as an alternative to mitigate such shortcomings. Implementing analog circuit functionality with highly digital circuits brings the benefit of scalability to analog functions. In this book we review the basic principles of time encoding, in particular applied to analog-to-digital converters (ADCs) based on Voltage-Controlled Oscillators (VCOs), one of the most popular time-encoding techniques to date. VCO-based ADCs are known since a long time [Hov97], [Nai00] but they become a mainstream research topic after Straayer and Perrott's 2008 paper [Str08]. Since then, many other advancements from academia and industry have been developed in application domains such as sensor readout, telecommunications, Internet of Things, automotive, or biomedical [Str08], [Ana16], [Tu17a], [Tu17b], [Qui20], [Van13], [Hua17], [Dan10], [Xin15], [Tay10a], [Opt10], [Wu19], [Muk19], [Bae19], [Wat03], [Ngu18a], [Wat18], [Taj14], [Yeo18], [Jia17a], [Jia17b], [Hua18]. Time encoding has also been used for other analog functions, especially by seizing the integrating properties of VCOs in state variable filters [Lee17], [Dro12].

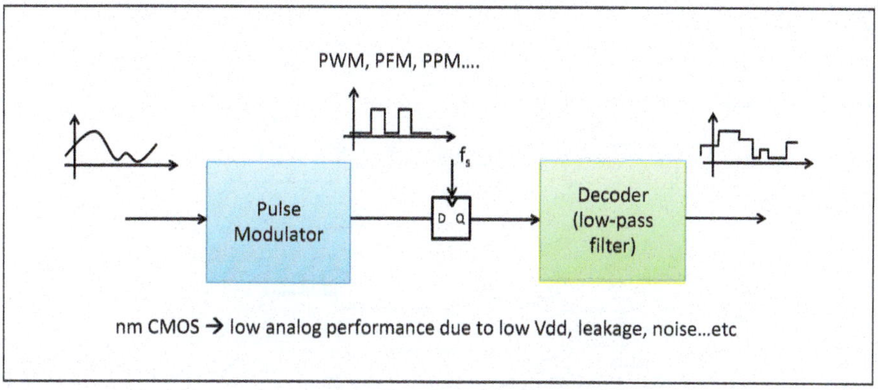

Fig. 1.2 Signal processing principles of time-encoding ADCs

Continuous-time signals convey information in their fluctuations in time. The fluctuation rate is determined by the signal bandwidth. Typical analog circuits represent signals with the minimum possible bandwidth, assuming that amplitude is properly transferred or processed because noise and distortion can be limited. However, it is possible to exchange amplitude accuracy by fluctuation speed, seizing the spectral spreading produced by some modulation techniques. Nanometer CMOS circuits are inefficient at amplitude accuracy but very good in representing time fluctuations due to the speed brought by very small devices in these technology nodes. Therefore, they may benefit from such spectral spreading techniques. This approach is what is commonly referred as "Time Encoding." The most robust signals in terms of amplitude representation are those that can take only two amplitude levels. These signals can be handled by digital circuits for which nanometer CMOS is optimized. To do so, we first need to describe a way to map amplitude-encoded signals into two-level time-encoded signals, and afterwards we need a new set of algorithmics that allow processing these time-encoded signals.

A solution for the first task, amplitude to time mapping, was developed at the beginning of digital communications with Pulse Modulations such as Pulse Width Modulation (PWM), Pulse Position Modulation (PPM), or Pulse Frequency Modulation (PFM). Pulse modulations encode a continuous band-limited analog signal into a two-level signal. Although perfect recovery of pulse modulations is mathematically possible [Laz04], it would require an ideally infinite time accuracy and complicated signal interpolation algorithms. In practice, pulse-encoded signals can be remapped into a smooth band-limited signal by simple low-pass filtering [Roz97] with sufficient quality.

The second task, finding new algorithms to process time-encoded signals, has received some attention [Tsi06] but is still not fully developed. Instead, time encoding has been used to implement data converter architectures that allow using standard discrete time signal processing algorithms. Time encoding enables hardware-efficient data converters because two-level signals can be sampled with digital flip-flops (see Fig. 1.2), replacing analog samplers, and because some pulse modulations convey the input signal in their low-pass spectrum. The preferred pulse modulations are PWM and PFM because both meet these two requirements.

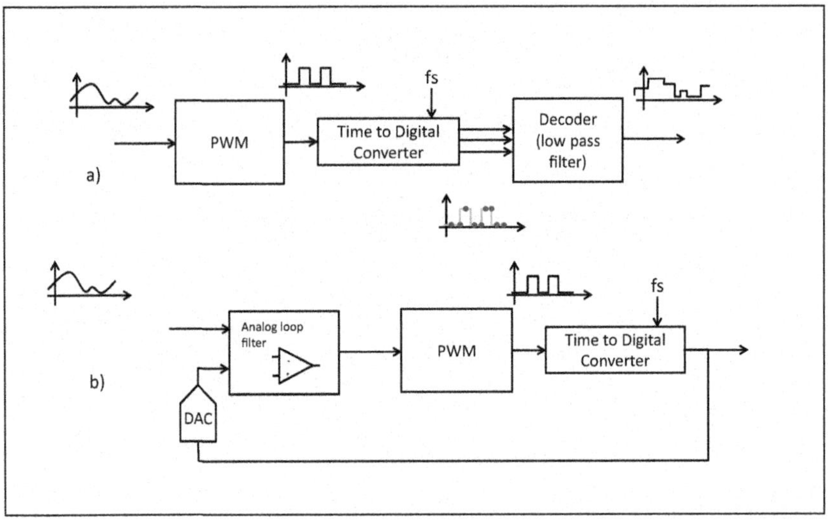

Fig. 1.3 Initial approaches to time encoding. (**a**) Open-loop PWM, (**b**) CT-$\Sigma\Delta M$ with PWM quantizer

Initial approaches to time-encoded data converters used PWM. The first converters used an analog Pulse Width Modulator followed by a Time-to-Digital Converter (TDC) [Roz97] (see Fig. 1.3a). The Pulse Width Modulator encodes the input analog signal into a two-level pulsed signal whose low-pass spectrum reproduces the input. The TDC is a purely digital circuit made of inverters that samples the PWM signal with a fine time resolution without requiring a high-frequency clock. The main shortcoming of this architecture is that a Pulse Width Modulator still requires a significant amount of high-precision analog functions such as integrators, comparators, or current sources. Therefore, the implementation advantages in nanometer CMOS are limited. Additionally, time errors introduced by sampling the PWM signal produce in-band quantization noise that severely limits the SNDR, restricting this technique to low-resolution ADCs.

As a possible solution to these limitations, the PWM-TDC circuit was used to replace the coarse quantizer of a conventional continuous-time Sigma-Delta modulator [Pre11], [Dha11] (see Fig. 1.3b). The results improved because the hardware requirements of the Pulse Width Modulator were relaxed and the TDC quantization noise was noise shaped. However, this was just a marginal improvement over conventional continuous-time Sigma-Delta modulators which still require high-performance operational amplifiers and precise RC time constants.

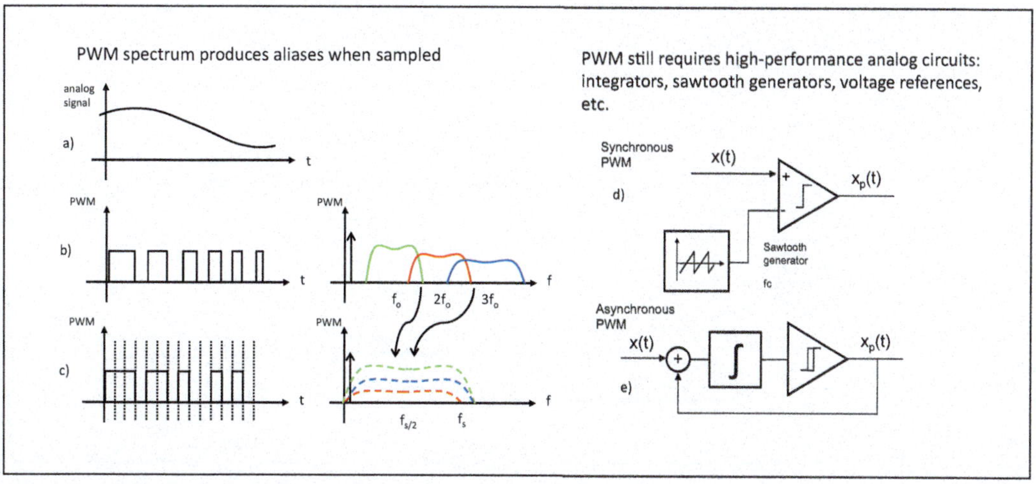

Fig. 1.4 Problems of PWM-based time-encoded systems

We are going to describe how PWM-based ADCs work from a system-level point of view. In the left side of Fig. 1.4, we have an analog band-limited signal (Fig. 1.4a) that is PWM-encoded in a periodic pulse train (Fig. 1.4b). The pulse repetition frequency is known as the carrier frequency. This PWM modulated signal is discrete in amplitude but continuous in time. Its spectrum has analytically been calculated for sinusoidal inputs and is represented at the right side of the corresponding time version in Fig. 1.4. The PWM spectrum is composed of the input signal (a low-frequency tone represented in black in Fig. 1.4b) and a set of sidebands centered in the frequency of the carrier and its harmonics (see green, red, and blue sidebands in Fig. 1.4b). The power of the sidebands decreases for higher harmonics while its bandwidth increases. With the PWM signal sampled, it becomes a binary discrete-time sequence. Therefore, its spectrum turns periodically around the sampling frequency and the harmonic sidebands are aliased to low frequencies. To avoid that harmonic sidebands invade the input signal bandwidth, a sampling frequency much higher that the input signal bandwidth is required. This results in a reduced power efficiency when comparing PWM-based ADCs with other architectures such as high-order Sigma-Delta modulators, which incorporate noise shaping.

On the other hand, the linearity and noise of a PWM-encoded signal depend on the PWM modulator. PWM modulators can be synchronous (see Fig. 1.4d) which use a periodic saw tooth signal and a comparator. Synchronous PWM modulators rely on the linearity and low noise of this saw tooth signal, typically generated with current sources, suffering from PVT variations, noise, and finite output resistance.

Other PWM modulators use a feedback self-oscillating architecture (see Fig. 1.4e) but produce both pulse width modulation and a signal-dependent frequency shift of the carrier, which complicates the sampling process.

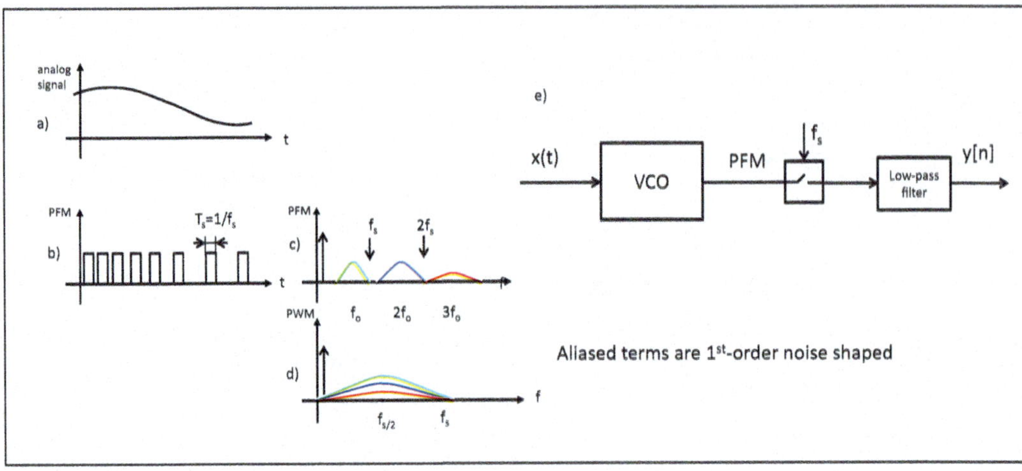

Fig. 1.5 A different pulse modulation: Pulse Frequency Encoding—VCO-ADC

As a difference to PWM, Pulse Frequency Modulation (PFM), which is used in VCO-ADCs, brings the benefit of noise-shaped aliasing errors after sampling. Assuming an input band-limited analog signal (see Fig. 1.5a), the output of a PFM modulator is composed of pulses with identical width T_s that are repeated often or sparsely depending on the input signal level (see Fig. 1.5b). A high input amplitude results in densely packed pulses occurring very fast. A low input amplitude results in very spaced pulses. At rest, the PFM modulator produces pulses with a constant carrier frequency. Therefore, the average value of the pulses over time follows the input amplitude. This property, also present in PWM, manifests very differently when we sample the PFM signal. The spectrum of PFM was calculated analytically for sinusoidal inputs [Fit47]. The spectrum is composed of the input signal at baseband (black tone in the frequency plot of Fig. 1.5c) and in the modulation sidebands centered at the carrier frequency and its harmonics (green, blue, and red in Fig. 1.5c) similarly as with the PWM. However, the modulation sidebands show periodic nulls in the spectrum at multiplies of the inverse of the pulse width $f_s = 1/T_s$. By sampling at precisely that frequency f_s, all modulation sideband nulls alias to DC and this results into noise-shaped quantization noise (Fig. 1.5d).

Although the PFM signal displayed in Fig. 1.5 is not immediately identifiable in a VCO-based ADC, the architecture of such converters uses an oscillator followed by a sampling flip-flop. Some digital post-processing circuit produces the equivalent PFM signal (see Fig. 1.5e) from the sampled VCO oscillation.

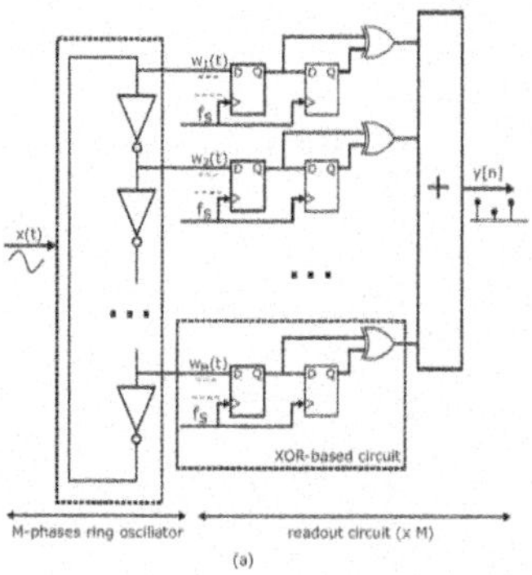

1. "Mostly digital" → Most building blocks are digital circuits (but some implemented as analog circuits)

2. "Time encoding" → Dynamic range not limited by supply, signal encoded in pulse timing

3.Shift current consumption from analog to digital domain

4. Multibit output; decimation filters can be very simple

5. Similar antialias as CT-ΣΔ but more jitter tolerant

Disadvantage: VCO distortion; Digital logic is notfor free -it may consume more than analog circuits.

VCO-ADCs can be a good candidate to integrate ADCs in a SoC

Fig. 1.6 VCO-ADC—implementation advantages

In spite of the advantages of PFM over PWM with respect to sampling, PFM would not be very practical if it was not because of a privileged implementation using mostly digital circuits. The VCO oscillator at the core of the VCO-ADC can be implemented with a ring oscillator (RO). A ring oscillator is a chain of digital inverters which sustain an oscillation due to their delay times. As long as the delay of an inverter depends on the supply voltage, it is easy to modulate the frequency of the RO with the input analog signal. Figure 1.6 shows the typical circuit used to sample the RO output square wave signals and decode a noise-shaped representation of the input signal out of the inverter outputs. This circuit has the advantage of producing simultaneously multiple phases of the oscillation. The circuit is also fully digital, revealing the interest in these architectures: in principle, they do not require any analog circuit to implement a full data converter. In addition, the dynamic range is not limited by the supply voltages, as signals are encoded in the frequency rather than the amplitude. This allows to shift most of the power consumption to digital circuits.

On the other hand, ring oscillators are mediocre VCOs with respect to linearity. This is the weak point of VCO-ADCs and most of the research and novel architectures today have as goal to improve the RO linearity without compromising its digital nature and low power consumption.

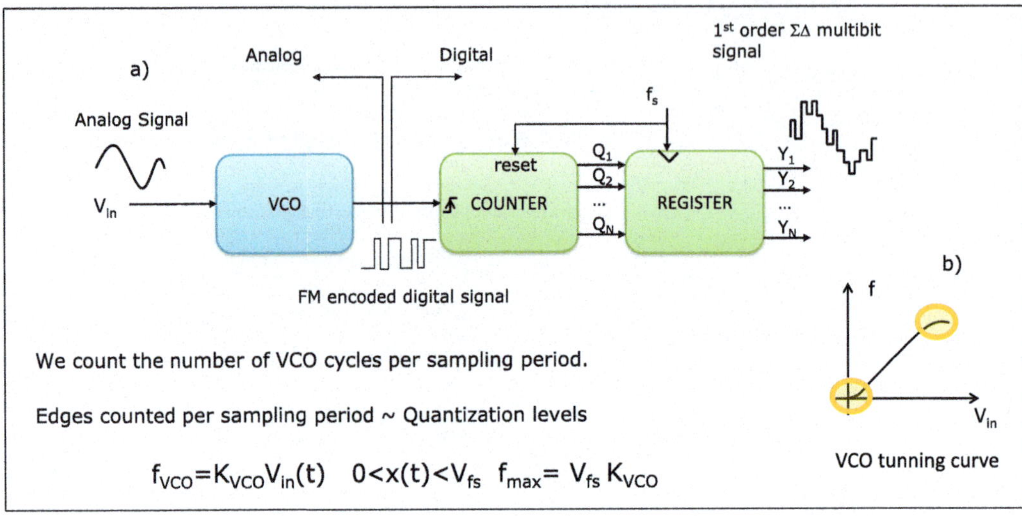

Fig. 1.7 What is a VCO-ADC?

We will explain the operation of a VCO-ADC from a signal processing perspective later, but first we are going to introduce VCO-ADCs with an intuitive explanation. Figure 1.7a shows the basic architecture of a VCO-ADC. It is composed of a VCO modulated by an input signal V_{in}, a digital counter that counts the pulses of the VCO and a register that captures the value in the counter. The register is triggered at the sampling frequency f_s. Every time we sample the counter, we reset its contents to zero and start a new count. Therefore, this system counts how many pulses the VCO produces in a sampling period $T_s = 1/f_s$. As long as the frequency of the VCO depends on V_{in} in an ideally linear way, the numbers sampled in the register are proportional to the frequency and as a consequence, to V_{in}. At the end, VCO-ADCs consist of counting the pulses of a VCO, no matter how complicated the circuit is. Obviously, the faster the VCO, more pulses will be counted and the resolution of the ADC will be better (i.e., more ADC bits). There is a further advantage in this architecture: as we will show later, the quantization error of this ADC is first-order noise shaped, and therefore oversampling brings also an SNR improvement.

The main problem of this architecture is the nonlinearity of the input voltage to VCO frequency relationship. An example of this relationship, named VCO tuning curve, is depicted in Fig. 1.7b. The VCO is operated at a central frequency f_0 corresponding to the middle of the input voltage range V_{in}. Nearby this operating point, the voltage to frequency conversion can be characterized by a linear relationship through a gain constant K_{vco}. However, for extreme values of V_{in} (see yellow shades in Fig. 1.7b), the VCO characteristic deviates from linear, which introduces distortion in the output signal.

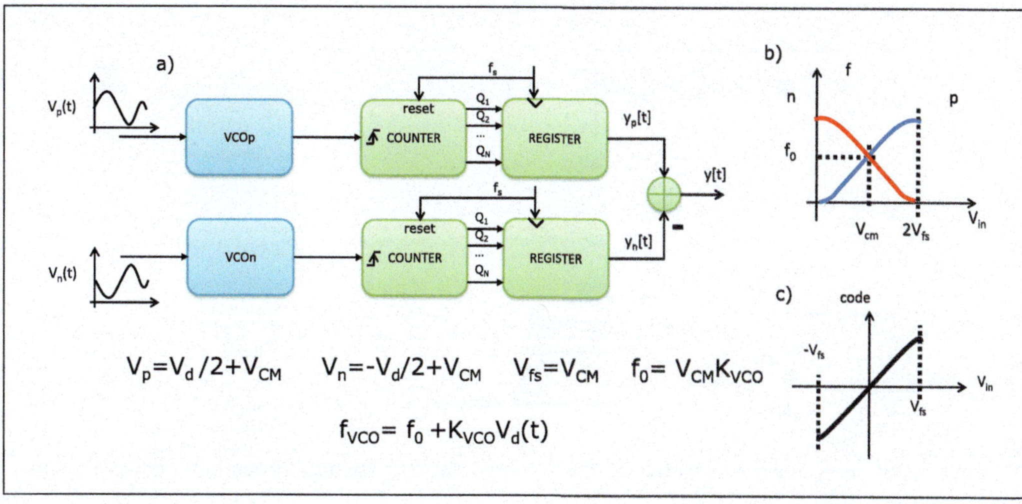

Fig. 1.8 Pseudo-differential configuration

A very common way to mitigate the VCO nonlinearity is using a pseudo-differential configuration. Figure 1.8a shows this configuration which consists in two twin paths that encode two analog signals V_p and V_n with the same architecture as in Fig. 1.7a. The analog signals are a differential representation of the input signal that can be split into common-mode V_{cm} and differential components V_d. Each of the VCOs will follow its tuning curve and if we assume matching between the two VCOs, they will oscillate to the same rest frequency f_o with a zero input ($V_p = V_n = V_{cm}$, $V_d = 0$) (see Fig. 1.8b). Both register outputs $y_p[n]$ and $y_n[n]$ are digitally subtracted afterwards, generating output sequence $y[n]$. In this way, the converter can represent signed magnitudes and inherits the even harmonic rejection property of differential systems, as shown in Fig. 1.8c. Most VCO-ADCs implemented to date use this architecture for its linearity enhancement and for its robustness to power-supply perturbations.

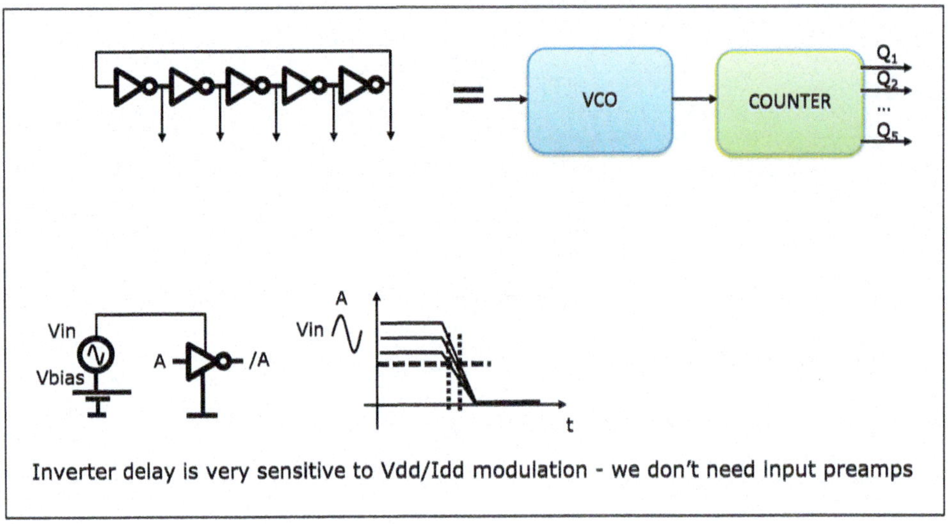

Fig. 1.9 VCO-ADCs advantage: ring oscillators. (**a**) Ring oscillators = VCO + Counter. (**b**) Ring oscillators = Amplifiers

In some VCO-ADCs, especially in sensor applications, the VCO-ADC is implemented as the counting architecture depicted in Fig. 1.7a, with a digital counter coupled with a single output VCO. However, we said that VCOs are better implemented with ring oscillators having multiple output phases, due to its digital nature: having multiple phases allows to pack in the same ring oscillator both the oscillating and counting functions (see Fig. 1.9a). This way, the power and area occupied by the counter are saved. As an additional advantage of ring oscillators, digital inverters are very sensitive circuits, as small V_{dd} variations around the proper bias voltage produce a significant propagation delay change (see Fig. 1.9b). This allows to implement ADCs with quantization steps on the order of a few μV without any preamplifier. This sensitivity has enabled the implementation of data converters in instrumentation and biomedical applications [Wat18], [Jia17a], [Jia17b], [Yeo18].

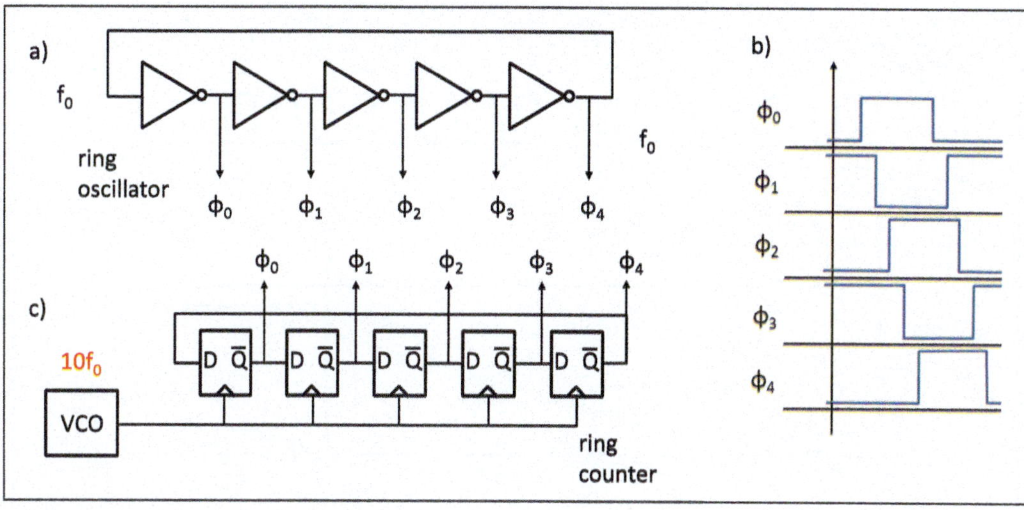

Fig. 1.10 Ring oscillator = Oscillator + Counter

As an example on why a ring oscillator can be seen as a VCO plus a counter, we show two equivalent circuits in Fig. 1.10. The circuit of Fig. 1.10a is a ring oscillator with five inverters. The output phases Φ_0–Φ_4 follow the pattern depicted in the chronogram (see Fig. 1.10b), where the transitions between edges are delayed by the propagation delay of the inverters T_d. Assuming a reference phase Φ_0, each phase output Φ_k will have a frequency f_0 and a phase shift 2πk/5 rad. The circuit of Fig. 1.10c shows a conventional oscillator having a frequency 10 times faster than the ring oscillator, $10f_0$, but a single output, like in the block diagram of Fig. 1.7. Now we connect a ring counter implemented with five D flip-flops as shown in Fig. 1.10c. By clocking this ring counter with the output of the VCO, we have at the ring counter outputs exactly the same sequence as with the ring oscillator in phases Φ_0–Φ_4. By observing the output sequence in the chronogram, we can see that it cycles through ten different states and therefore can be used as a counter, albeit it does not provide a binary output. This will not be a problem but rather an advantage, lets calculate the count increment in a sampling period with a simple array of XOR gates.

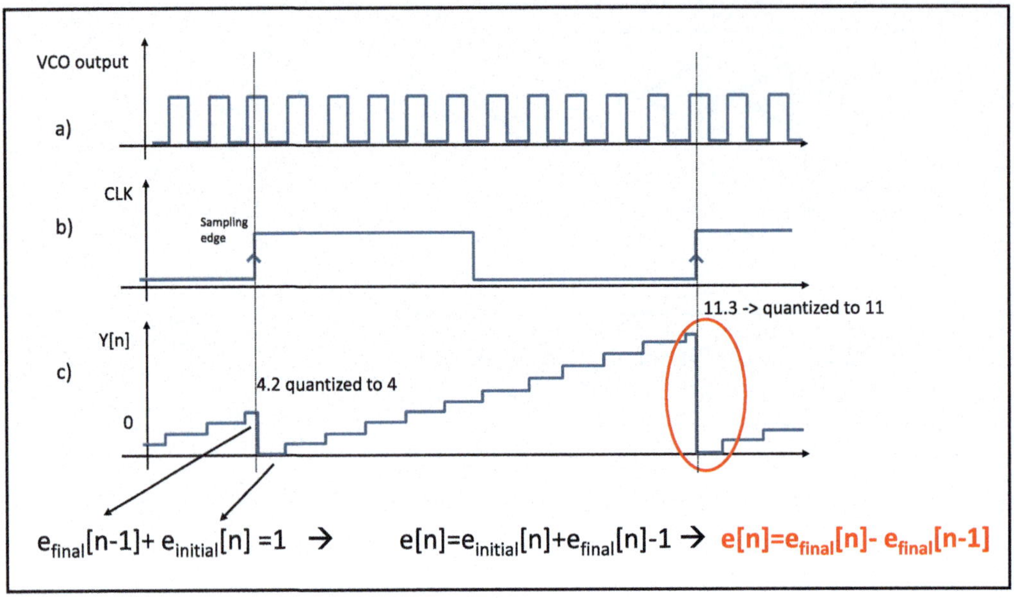

Fig. 1.11 VCO-ADCs also do first-order noise shaping

We have shown in Fig. 1.7 a simple model of a VCO-ADC based on counting the VCO pulses and we stated there that this system can be seen as a quantizer with spectrally shaped quantization error. We will try to justify this with another intuitive explanation, by investigating the different signals present in the block diagram of Fig. 1.7, shown in the chronogram of Fig. 1.11. The output of the VCO is the square waveform in the plot of Fig. 1.11a. We are going to sample the count of the counter of Fig. 1.7 with the rising edge of a clock of frequency f_s, shown in the plot of Fig. 1.11b. The sampling clock is totally asynchronous to the VCO oscillation and therefore, the sampling edges can happen anywhere between the VCO pulses. The amount stored in the counter is represented as the staircase plot of Fig. 1.11c. At every sampling edge, we capture the counter value and reset the counter to zero. This produces a timing error because the counter value is an integer number but the exact amount of pulses between sampling edges has an initial time error e_{initial} and a final error e_{final} (see Fig. 1.11c). These errors can be seen as random variables given that the sampling clock and the VCO output are independent signals. Obviously, the final error of sample $[n-1]$, $e_{\text{final}}[n-1]$, plus the initial error of the next sample n, $e_{\text{initial}}[n]$, add to 1 VCO pulse. Therefore, we can write the quantization error sequence as the first difference of a random sequence uniformly distributed between 0 and 1, whose spectrum will have a null at DC and shows first-order noise shaping as a consequence.

maximum # of counts per sampling period

OSR

$$SQNR_{max}(dB)=6\log_2(2f_0/f_s)-5.17+1.76+30\log_{10}(f_s/2ABW)$$

- If we double the rest oscillator frequency f_0, we improve SQNR by 6dB, (1 bit)
- If we also double the sampling frequency f_s, we improve SQNR by 9dB, (1.5bit)

Fig. 1.12 Approximate SQNR formula (first-order ΣΔ)

Now that we qualitatively have shown the operation of VCO-ADCs we are going to make some calculations on the performance that can be obtained with the VCO-ADC of Fig. 1.7 with respect to the main design parameters such as sampling frequency f_s and oscillation frequency f_0. The fact that quantization noise is first-order shaped allows us to borrow the Signal to Quantization Noise (SQNR) formula from Sigma-Delta modulation theory. In spite of this, the formal equivalence between a VCO-ADC and a first-order Sigma-Delta modulator will be established later on. In this SQNR formula, the number of bits of the quantizer is equivalent to the maximum count that can be reached in the counter, expressed as the required bit width of the counter.

The maximum count is the maximum frequency of the VCO divided by the sampling frequency. We assume that the VCO oscillates at frequency f_0 at rest and at $2f_0$ with the full-scale input. The oversampling ratio OSR can be defined, as like in standard in Sigma-Delta modulation, as the ratio between sampling frequency and twice the input signal bandwidth ABW. The resulting formula is shown in Fig. 1.12. Here we see a trade-off between sampling frequency and oscillator rest frequency. On the one hand, if we lower the sampling frequency, the bit width of the counter increases. On the other hand, if we increase the sampling frequency, the oversampling factor increases. Both of these increments result in a SQNR improvement, which is a bit a confusing conclusion. As a rule of thumb, we can say that if we double the VCO oscillation frequency, the SQNR improves by 6 dB. But if we also double the sampling rate at the same time, we improve the SQNR by 9 dB, which defines more clearly the system behavior as a first order system.

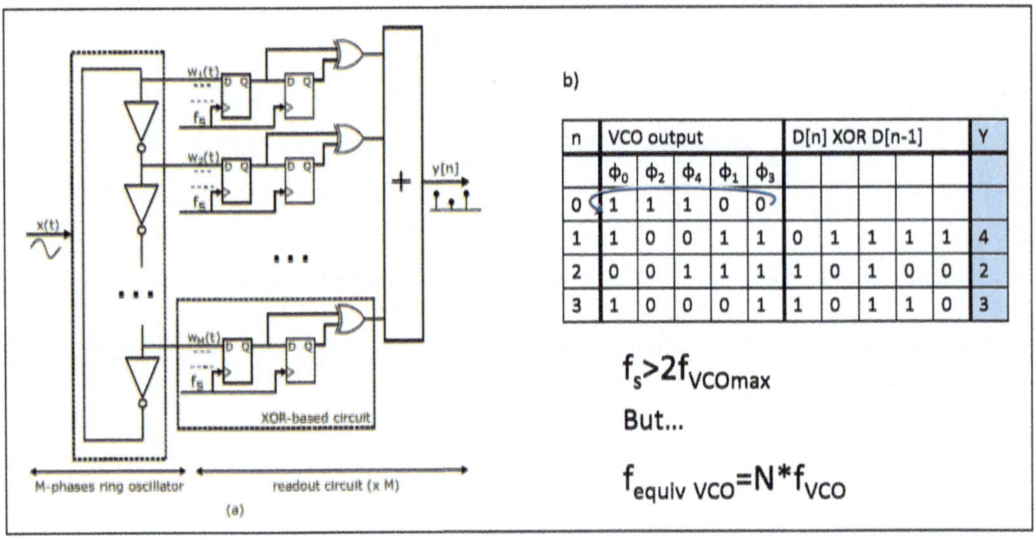

Fig. 1.13 A hardware-efficient implementation for the counter

We have shown that a ring oscillator (see Fig. 1.10) can be seen as the combination of a VCO followed by a counter. However, the count is not a binary number but a sequence of cyclic states. The code produced by the counter can be seen as a group of ones and zeroes that shift one position at a time. To visualize this more clearly, we can simply reorder the five phases of the oscillator of Fig. 1.10 in a set of odd and even phases (see Fig. 1.13b). The number of shifts is proportional to the count that we would expect in the equivalent counter. Therefore, we can calculate the count by calculating the difference between the position of the ring oscillator at the beginning ($t = (n - 1)T_s$) and at the end ($t = nT_s$) of each sampling period. As long as the ring oscillator does not rotate a complete cycle in one sampling period (i.e., f_s is larger than twice the maximum VCO frequency), the number of state transitions can be calculated as the bitwise XOR of each ring oscillator phase. The table of Fig. 1.13b shows an example of a sequence of codes obtained from the circuit of Fig. 1.10 and its decimal equivalent number.

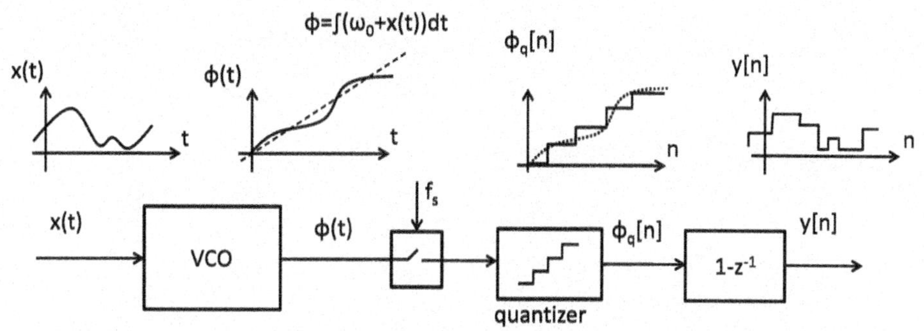

$$\phi_q[n] = w_0 n T_s + k_{VCO} \int_{-\infty}^{nT_s} x(\tau) d\tau + q[n]$$

$$y[n] = \phi[n] - \phi[n-1]$$
$$= q[n] - q[n-1] + w_0 T_s + k_{VCO} \int_{(n-1)T_s}^{nT_s} x(\tau) d\tau$$

Fig. 1.14 The phase-referenced model

We have assumed the equivalence between a VCO-ADC and a first-order Sigma-Delta modulator in order to estimate the SQNR. This equivalence can be supported by some analytical calculations that were established at the beginning of VCO-ADC analysis, for instance, in [Hov97], [Tay10a], [Kim10]. In Fig. 1.14 we have depicted a VCO at system level with signal $x(t)$ as input, and as output variable, we have used the phase $\Phi(t)$. The phase, by definition, is the integral of the frequency, and therefore is proportional to the integral of the input signal $x(t)$ plus a linearly growing term representing the rest frequency f_0. This ever growing phase signal $\varphi(t)$ is represented on top of the block diagram of Fig. 1.14. Obviously, the output of a real VCO (like a ring oscillator) is a square wave that has rising and falling edges coincident with the even or odd phase crossing by multiples of π. Therefore, what we really observe from the phase is a quantized version with quantization steps of π radians. If we sample the quantized phase of the VCO, we obtain a staircase plot (see Fig. 1.14). The sampled and quantized phase signal is a discrete sequence growing to infinity, but we are interested in the phase increment in a sampling period. We can compute the first difference of the phase and then we will obtain a plot that is exactly the plot of Fig. 1.11, obtained with the physical VCO plus counter setup of Fig. 1.7. The equations tell us two important facts: the quantization noise is in effect first-order shaped and the input signal is integrated during every sampling period, which can be seen in the frequency domain as passing through a low-pass sinc-shaped filter with nulls at the sampling frequency. This sinc filter can be defined as the Signal Transfer Function (STF) of a VCO-ADC, following the parallelism with Sigma-Delta modulation theory.

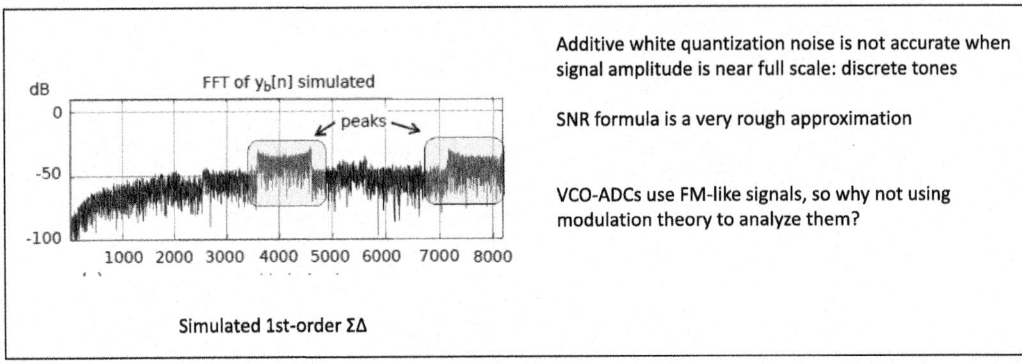

Fig. 1.15 But quantization noise of first-order ΣΔ is not "white"...

The explanation theory on the phase of the VCO establishes a bridge between Sigma-Delta modulation and VCO-ADCs, showing the equivalence with a first-order Sigma-Delta modulator. However, the first-order modulator is the case where the white noise assumption for quantization noise fails more clearly. If we simulate a first-order modulator, it is easy to see that the spectrum of the quantization noise is far from white at some points (see the peaks in the simulation of Fig. 1.15). Also, the white noise model of Sigma-Delta modulation does not explain clearly the effect of the VCO center frequency in the SQNR, but simulations show a strong influence, same as in a Sigma-Delta modulator with a nonzero DC input. These facts and the connection that VCO-ADCs should have with frequency modulation have triggered a further research on an improved explanation of VCO-ADC behavior. It has been found in [Gut18a] that VCO-ADCs can be better explained using Pulse Frequency Modulation, as stated in the introduction. We are going to explain briefly this analysis, which confirms the results of Sigma-Delta theory and provides an analytical description of VCO-ADC quantization noise with sinusoidal inputs.

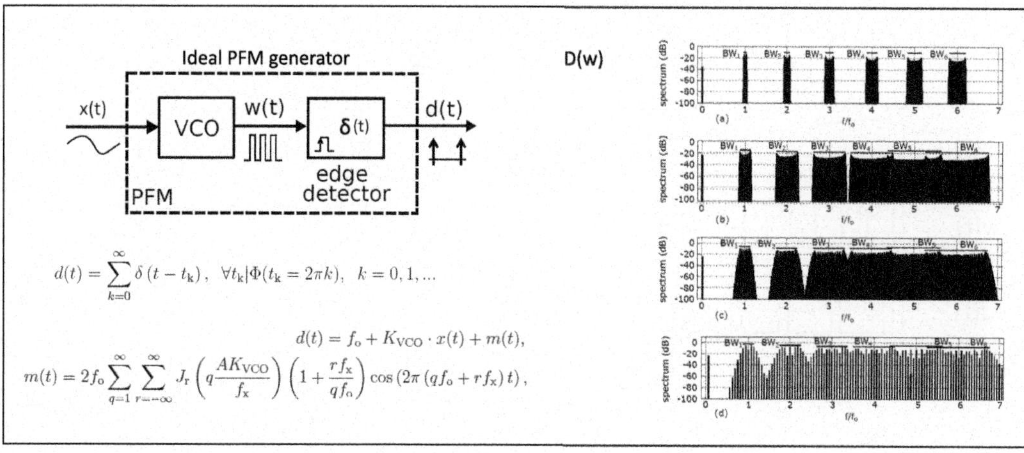

Fig. 1.16 What is Pulse Frequency Modulation (PFM)?

First of all, we are going to introduce Pulse Frequency Modulation (PFM). In pulse frequency modulation we generate identical pulses of an arbitrary shape that are repeated with a frequency proportional to the input signal. A system encoding a signal with PFM is a pulse frequency modulator. Figure 1.16a shows an example of a pulse frequency modulator at system level: it consists of a linear VCO producing a square wave $w(t)$ that is frequency modulated by input $x(t)$ and followed by an edge detector. The edge detector produces a train of Dirac delta functions $d(t)$ located at the edges of the VCO square wave. This is the simplest form of a pulse frequency modulator, because if we want the PFM signal to use a particular pulse shape, it is sufficient to place a linear filter having the desired pulse shape as impulse response. Whenever we apply a Dirac delta pulse at the input of this linear filter, the output will be the desired shaped pulse located at that precise time instant.

If the input $x(t)$ is a sine-wave, output $d(t)$ can be expanded as a Fourier series. This Fourier series was calculated first in [Fit47] for square pulses and afterwards in [Bay68] for the generic Dirac delta case. This Fourier series has two components, one conveying the input signal $x(t)$ plus an offset f_0 depending on the rest oscillation frequency. Another modulation component $m(t)$ consists of sidebands with multiple tones spaced the frequency of the input signal $x(t)$ and repeated for all harmonics of the rest frequency f_0. The harmonic sidebands are spectrally shaped by a set of Bessel functions. In general, the harmonic sidebands increase in bandwidth but decrease in amplitude proportionally to the harmonic order. Also, the harmonic sideband bandwidth depends on the amplitude of the modulating input $x(t)$. Typically the frequency of the input $x(t)$ is much smaller than that of f_0 and the harmonic sidebands are far from the input signal in the spectrum. This allows to recover the input signal by low-pass filtering the signal $d(t)$.

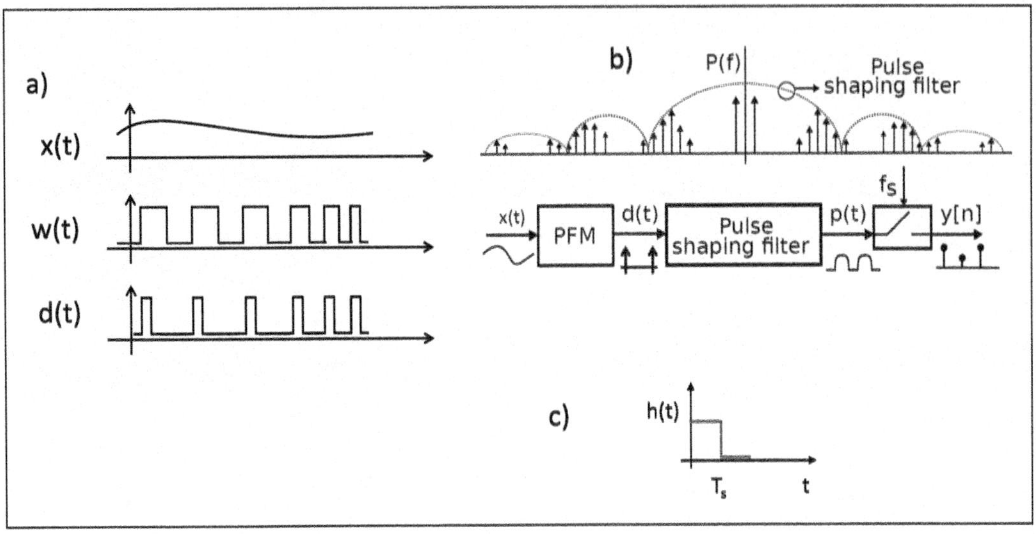

Fig. 1.17 Pulse Frequency Modulation with a square pulse

The case where the shaping pulse of a pulse frequency modulator is a square pulse is of special interest. In the chronogram of Fig. 1.17a, we have depicted the input waveform $x(t)$ of a PFM modulator (see Fig. 1.17b) as a smooth, slowly varying input. Signal $w(t)$ in Fig. 1.17a represents the VCO output and is a square wave whose instantaneous frequency follows the trend dictated by $x(t)$. We assume now that we place the linear filter with square impulse response $h(t)$ of duration T_s shown in Fig. 1.17c. Then the output of the shaping filter ($p(t)$ in Fig. 1.17b) will reproduce a square pulse at the rising edges of the signal $w(t)$. We have chosen the rising edges for simplicity.

Looking at how the square pulse modifies the spectrum of $d(t)$ we can state that the Fourier series of $d(t)$ is multiplied by the transfer function of the square pulse $h(t)$ as long as the filtering operation is linear. The Fourier transform of a square pulse $h(t)$ is a sinc function with unity gain at DC and nulls at every multiply of $1/T_s$. Therefore, the modulation sidebands $m(t)$ expressed in the equation of Fig. 1.16 are spectrally shaped by the sinc, as shown in the plot of Fig. 1.17b.

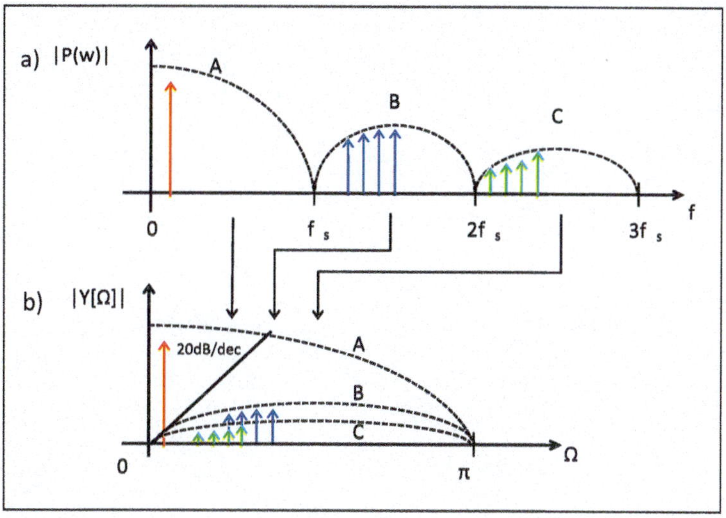

Fig. 1.18 Sampling process

Now, let us assume we sample signal $p(t)$ of Fig. 1.17b with a sampler exactly every T_s seconds. All nulls of the spectrum of $p(t)$ will alias to DC and therefore, all the modulation sidebands in the spectrum of $d(t)$, no matter where are they located (i.e., regardless of f_0) will alias to the Nyquist zone between 0 and $f_s/2$, shaped by the first-order nulls of the sinc function. We have depicted in Fig. 1.18a the same plot of Fig. 1.17b but marking with colors the different harmonics (blue first harmonic at f_0, green second harmonic at $2f_0$, etc.) and the input signal in red. After sampling, the input signal is not attenuated because it lies within the first lobe of the sinc. As long as $x(t)$ is oversampled, the input tone (in red) is affected by the gain of the sinc close to DC where it is maximum. However the aliases of the modulation sidebands follow the amplitude dictated by the sinc lobes above f_s and have zero amplitude at DC. This is an alternative explanation to the phase-domain model of Fig. 1.14 or the intuitive approach of Fig. 1.11. The difference is that in Figs. 1.11 and 1.14 we are assuming that the quantization error is "white" without any further investigation, but in the PFM representation we can analytically calculate the exact formula predicting the spectrum of the aliases from the PFM Fourier series. This equation (see Fig. 1.16) was calculated in [Her15], [Gut18a]. The reason why the white noise approximation works in some cases is that, for large input signals, the modulation sidebands $m(t)$ overlap, forming a multitone signal whose spectrum resembles a white noise (see $D(w)$ in Fig. 1.16).

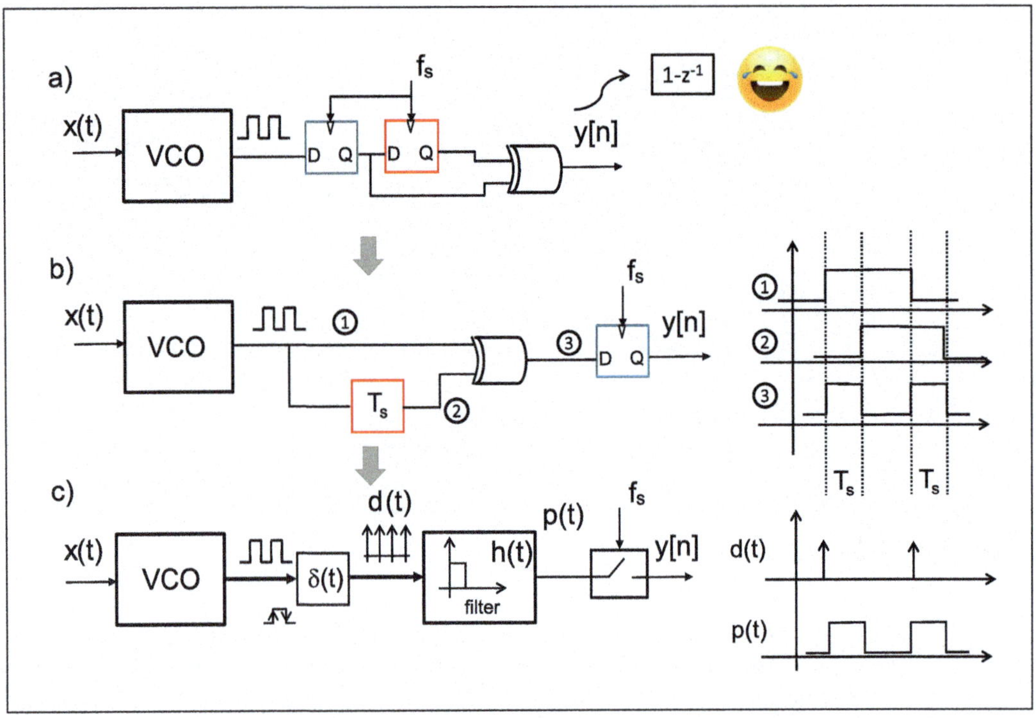

Fig. 1.19 Equivalence between a VCO-ADC with XOR decoder and a sampled Pulse Frequency Modulator

After showing the similarity of a sampled pulse frequency modulator and a first-order Sigma-Delta modulator, we may question what does this have to do with VCO-ADCs? We must now establish a relationship between the building block of Fig. 1.17b and the hardware implementation of Figs. 1.7 and 1.13. The fact is that this relationship is very straightforward but it has been hiding in the way circuits are drawn for some time. Consider the VCO-ADC having a single output like in Fig. 1.19a, The square wave produced by the VCO is sampled by a D flip-flop, and then the XOR function between present and past samples is computed, generating a single-bit, first-order shaped signal $y[n]$. This is the conventional circuit implemented in many papers, and the XOR function plus the sampling and delay flip-flops are attributed a first-order difference function $(1 - z^{-1})$. However this XOR is often confused with the first-order difference that the phase-referenced model applies to the phase of a VCO, as explained in Fig. 1.14.

Now let us rearrange the circuit in a manner that produces exactly the same sequence $y[n]$ as Fig. 1.19a that resembles the pulse frequency modulator of Fig. 1.17b. In the circuit of Fig. 1.19b we have moved the delaying flip-flop and the XOR function to the input of the sampler, replacing the flip-flop by a continuous time delay of T_s seconds. Clearly, such circuit will produce the same signal as the circuit of Fig. 1.19a, as shown in the chronogram on the right, that displays the internal nodes of this new circuit. Signal ③ in the chronogram happens to be exactly the signal $p(t)$ in the block diagram of a pulse frequency modulator with square shaping pulses, as reproduced in Fig. 1.19c. This shows the total equivalence between the PFM representation and the standard VCO-ADC architecture using a XOR and a delaying flip-flop.

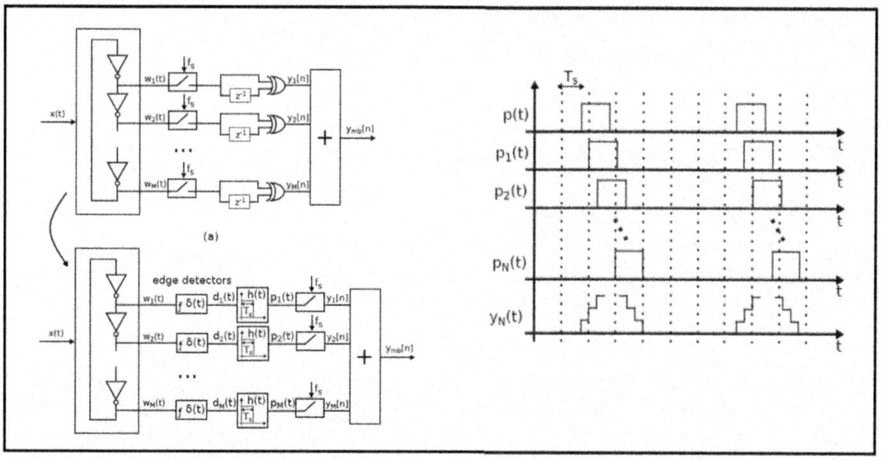

Fig. 1.20 Pulse Frequency Modulation applied to a Multi-Phase Ring Oscillator

Although we have established the equivalence for a single-phase VCO and a pulse frequency modulator in the previous figure, a practical VCO-ADC is in many cases constructed with a ring oscillator having many output phases. Therefore an extension to this more practical case is needed. This extension can easily be established by observing the typical decoding circuit of a ring oscillator based VCO-ADC, reproduced in Fig. 1.20a, which was introduced in Fig. 1.13. The XOR plus flip-flop circuit is repeated for every VCO output phase. Therefore, we can apply the same transformation of Fig. 1.19 in this case and assume that every output phase of a ring oscillator is a PFM-modulated stream of square pulses that combine together into a multi-level staircase waveform. The Fourier series of the output in this case can be calculated as a pulse frequency modulator operating at N^*f_0, with N oscillator taps, followed by a shaping pulse filter of length T_s. As long as N^*f_0 is much larger than f_s, the shaping filter output produces overlapping pulses creating the typical multi-level output of a ring oscillator based VCO-ADC.

Chapter 2
Quantization Noise, Thermal Noise, Flicker Noise, Phase Noise, and Clock Jitter in VCO-ADCs

G. Gielen et al., *Time-encoding VCO-ADCs for Integrated Systems-on-Chip*,
https://doi.org/10.1007/978-3-030-88067-5_2

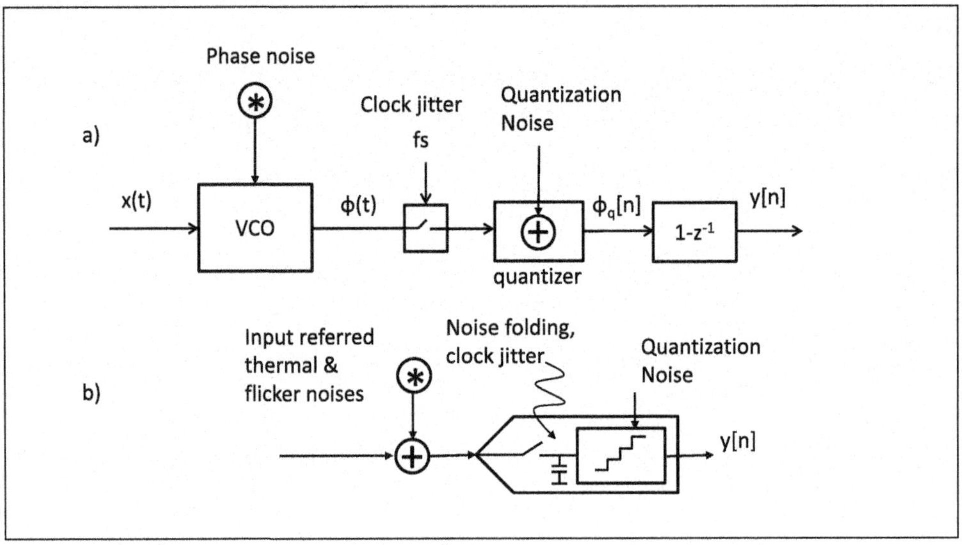

Fig. 2.1 Main noise sources in VCO-ADCs compared to SAR/ΣΔ

After describing the signal processing foundations of VCO-ADCs, we are going to discuss the main circuit implementation impairments that affect VCO-ADCs. Figure 2.1 compares the noise sources in a VCO-ADC and in a sampled ADC such as a switched-capacitor Sigma-Delta modulator or a SAR ADC.

In a VCO-ADC, we encode the input signal into the edge positions of a square wave using an oscillator. Oscillators, however, are affected by phase noise, which introduces timing errors in the oscillator output. As a consequence, the performance of a VCO-ADC is affected by the VCO phase noise (see Fig. 2.1a) in a similar manner as thermal and flicker noise affect the performance of a sampled ADC (see Fig. 2.1b). At the end, phase noise is produced as well by thermal and flicker noises in the electronic elements of the VCO, but the mechanism for which noise couples to the output data is a bit more intricate than in a sampled ADC. We will address this mechanism in detail.

Another noise source that affects both sampling and VCO-based ADCs is clock jitter. Such sampling clock jitter affects each ADC variety with some differences. For instance, VCO-ADCs can tolerate clock jitter by using a pseudo-differential architecture. We will also address in detail how to dimension the VCO-ADC to meet the required quantization noise. Finally, distortion is the other large impairment affecting VCO-ADCs. This will be the subject of the foregoing parts of this book.

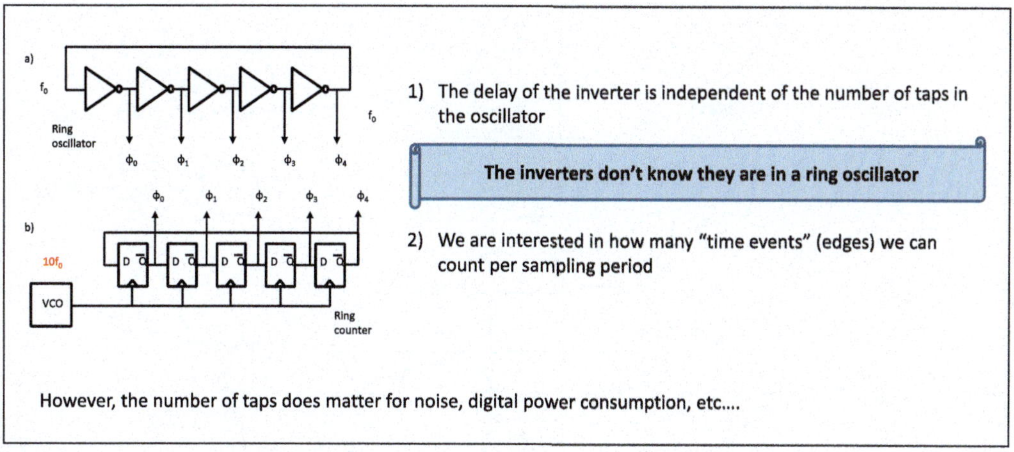

Fig. 2.2 A few words about ring oscillators first...

Before we describe the design process of a VCO-ADC from the quantization noise perspective, we will focus on the use of ring oscillators in VCO-ADCs. The oscillation frequency required in a VCO-ADC together with the sampling rate determines the SQNR (see Fig. 1.12). Once we have defined such frequency, one question that pops up immediately is how many phases do we need in a ring oscillator to implement the VCO. The answer is very simple: any. What really matters in a VCO-ADC is the minimum delay required in one of the inverters of the ring oscillator, as this defines the "effective frequency" of the oscillator. Recall that in Fig. 1.10 we defined the equivalence between a 5 tap ring oscillator and a digital counter driven by a 10 times faster oscillator. The equation in Fig. 1.12 refers to a single-phase VCO. Then we can apply the equation to find the effective frequency and later on decide how many taps we will use, trading off taps for ring oscillation frequency. In a sense, the inverters in the ring "don't know they are in a ring oscillator," just that they delay a digital signal (Fig. 2.2).

However, the number of taps in the ring oscillator does affect other factors such as flicker noise performance or the complexity and power consumption of the decoding digital logic. Therefore, deciding the number of taps of the ring oscillator is something we can initially delay when defining the quantization noise requirements.

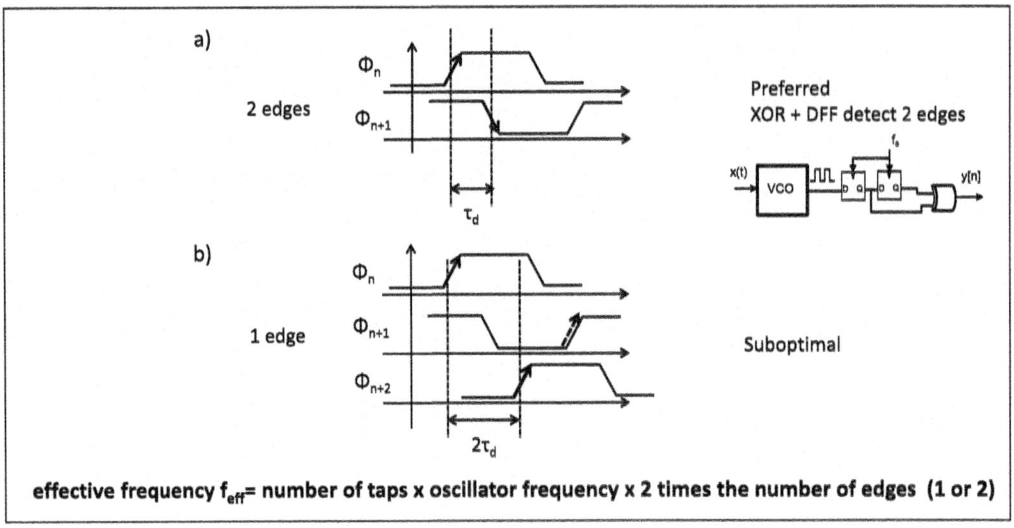

Fig. 2.3 The effective frequency

The definition of the "effective frequency" of a VCO depends on the digital architecture that samples and processes the oscillator output phases. In the end, we are counting time events, namely the edges of an oscillating square wave. We can implement decoding logic that counts either rising or falling edges, as happens with conventional digital counters (see Fig. 1.7), but we would double the resolution if we could count both the rising and falling edges at the same time. The circuit based on a XOR gate and a delaying flip-flop is able to detect both edge polarities, for instance.

In Fig. 2.3, we analyze the signals in consecutive output phases of a ring oscillator. Figure 2.3a compares two consecutive outputs ϕ_n and ϕ_{n+1}. As we can see, the delay between a rising edge and the next falling edge is τ_d, the propagation delay of one inverter. However, if we consider only the rising edges, Fig. 2.3b shows that the next rising edge after phase ϕ_n happens in phase ϕ_{n+2}, $2\tau_d$ later on, revealing less time resolution which translates into less amplitude resolution at the input of the ADC than in the dual-edge case.

One may think that the circuit of Fig. 2.3a is then the optimal to implement the decoding logic, but some other considerations apply. For instance, if we want to achieve a large dynamic range in a sensor application, it may be interesting to run the VCO at a frequency much higher than the sampling frequency, and the XOR circuit cannot be used because it is restricted to oscillation frequencies below the sampling rate. In this case, a digital counter can be a better choice.

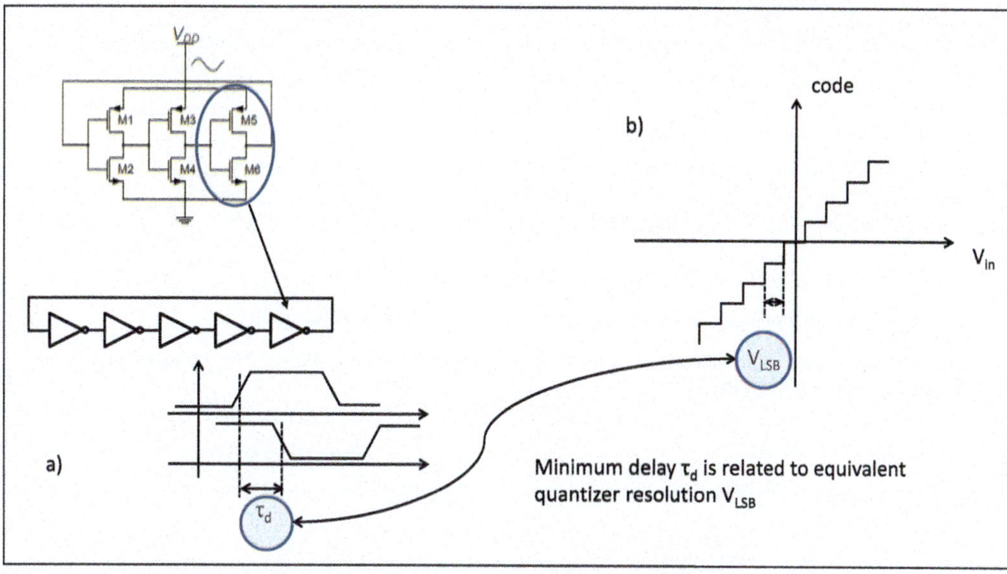

Fig. 2.4 Equivalence between a ring oscillator and a conventional quantizer

In a sense, we can think in the smallest delay in a ring oscillator inverter ($\tau_{d,min}$ in Fig. 2.4a) to be equivalent to the V_{LSB} of a quantizer (Fig. 2.4b). To design the VCO, we can start with a guess on the size of the transistors in the inverters by defining this minimum delay ($\tau_{d,min} \sim 1/2f_{eff}$) and carrying out a transient simulation of the inverter to check if we can meet that delay. Then, we will have to trade-off current consumption for transistor area to meet the required thermal and flicker noise specifications yet keeping the minimum delay. This allows to start the design regardless of the number of oscillator taps. On the other hand, using all the inverters in the ring oscillator with all the delay taps that we intend to use may lead to lengthy simulations and we can save time by designing and simulating just a single delay cell initially.

1) Define desired SQNR some dB above specs to allow for termal and flicker noises and distortion to dominate SNR.

2) Define sampling frequency f_s, OSR and if the digital postprocessing will use 1 edge or 2 edges

3) Calculate the required effective frequency

$$SQNR=6\log_2(2\,f_{eff}/f_s)-5.17+1.76+30\log_{10}(OSR)$$

$$f_{eff}=2\text{^}((SQNR+5.17-1.75-30\log10(OSR))/6-1)$$

f_{eff}= Number of taps (N) x oscillator rest frequency (f_0) x 2 number of edges

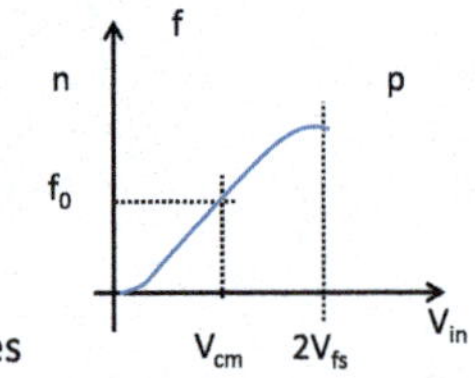

4) In a N-tap ring oscillator with rest frequency f_0, we have $f_0=f_{eff}/N/\text{Number of edges}$

Now we need to do a thermal/flicker noise analysis, digital power estimation....After all this, we set N, but now we can start designing our inverters:

$$\tau_d=1/2f_{eff}$$

Fig. 2.5 Quantization noise in ring-oscillator VCO-ADCs

We are now going to describe a possible strategy to design a VCO-ADC at system level. As a first step, we will define what is our target SQNR, providing some dB of margin in excess to the thermal noise level we require. This is a standard practice in ADC design that helps to optimizing power consumption.

The second step is to define if our digital decoding logic detects both rising and falling edge polarities (as with the XOR and flip-flop decoding circuit) or if we are going to use a counting circuit to estimate the VCO frequency that can detect only one edge polarity. In most cases we will detect the two edge polarities as it is the most power-efficient solution.

The third step is to define the effective rest frequency of the VCO. The effective rest frequency of the VCO (f_{eff}) is the frequency produced by a single-phase oscillator when we apply the input signal common mode, V_{CM} (see Fig. 1.8b). We will assume that the VCO frequency excursion ranges between zero and $2f_{eff}$, although in practice we will never reach neither zero nor $2f_{eff}$ to avoid saturation. In most cases, the oversampling ratio OSR and the sampling frequency f_s will be given by system-level considerations and therefore, we can consider them as inputs to our calculation. The effective frequency can approximately be calculated using the equation in Fig. 1.12, reproduced in Fig. 2.5.

As a fourth step we define the minimum delay that the inverters in our ring oscillator require as $\tau_d = 1/2f_{eff}$. The ring oscillator can have as many taps as we want and the decision is defined by flicker noise and digital power considerations.

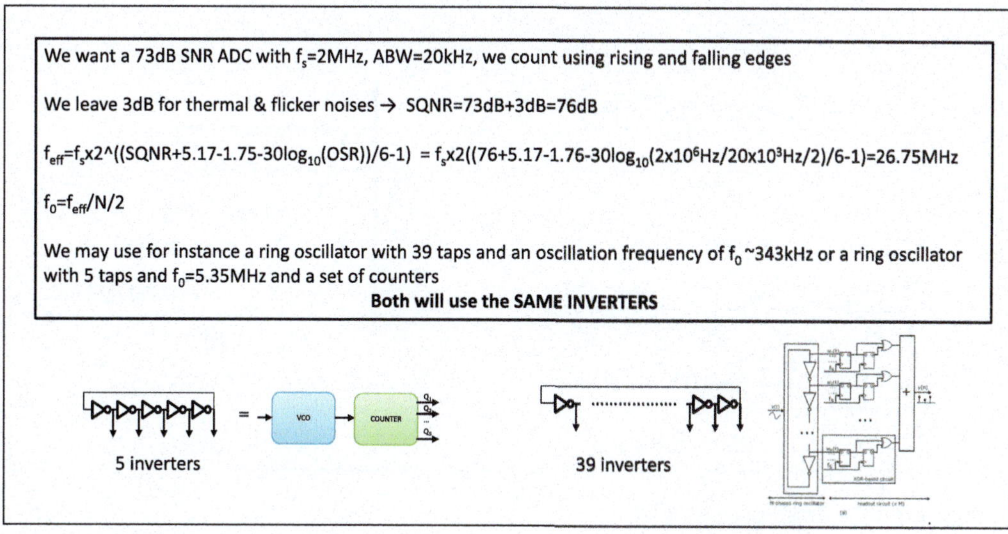

Fig. 2.6 Example of ring-oscillator design

To clarify the proposed design methodology, we are going to show a design example. Let us assume that we have 73 dB as target SNR of our ADC. The sampling frequency is 2 MHz and our ADC is for an audio application with an analog input bandwidth (ABW) of 20 kHz. We allow 3 dB of margin for quantization noise such that SNR is dominated by thermal noise instead of ADC quantization noise. Then, the required SQNR is 73 dB + 3 dB = 76 dB. With these data, and assuming two-edge detection, the effective frequency results in $f_{eff} = 26.75$ MHz. The minimum inverter delay will be 18 ns (Fig. 2.6).

Now we show two extreme implementations of the ADC, both of them having the same 76 dB of quantization noise:

(a) We decide to use $N = 5$ inverters. Then, the oscillation frequency of the ring oscillator should be 5.35 MHz. The VCO will run faster than the sampling frequency (2 MHz). Then, an array of delaying flip-flops and XOR gates (as in Fig. 1.13) cannot be used and we would need to use an alternative counting architecture (see, for instance, [Qui20], [Dan10], [Wat03], [Unn17]).
(b) We decide to use 39 inverters, then the oscillation frequency would be 343 kHz. We must not be misled by this apparently low frequency because the inverters are exactly the same as in case (a) and the power consumed by the VCO is also very similar (only switching consumes power and we have the same number of inverter switching per second). In this case we may use the XOR—flip-flop decoding logic and, as an advantage, the flicker noise will be 9 dB lower than in the case of 5 inverters (more on this later on). However, we would need to implement 78 D-flip-flops that are clocked at 2 MHz and after the decoding logic we would need a 39-to-6 bit thermometer decoder to convert the code coming out of the XOR gates into a binary number.

These are the trade-offs in the design of the ring oscillator.

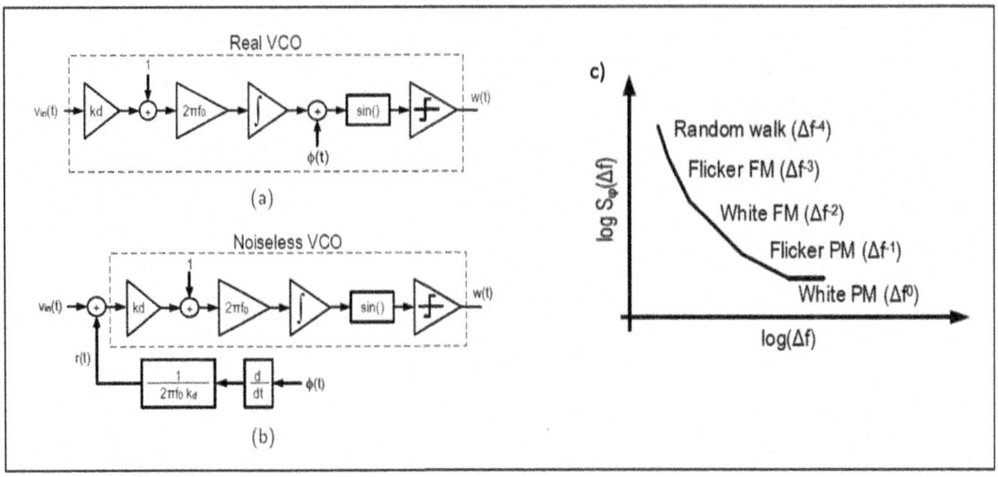

Fig. 2.7 Input-referred phase noise

In Fig. 2.1 we have observed that thermal and flicker noise manifest in VCO-ADCs as VCO phase noise. Analyzing the effect of phase noise in the SNR directly in the VCO is not very straightforward. On the other hand, a transient simulation with noise results in long simulations and a time-consuming task given the oversampled nature of VCO-ADCs. A better way to analyze the effect of phase noise is referring it to the input of the VCO and assuming the VCO is ideal. This analysis method gives us an ADC model that is similar to that of conventional sampled ADCs with noise sources that combine additively with the input signal.

The most common model for phase noise is the Leeson model (see Fig. 2.7c) which identifies several frequency bands where phase noise is dominated by different physical phenomena. Thermal noise imposes a zone where the phase noise depends inversely to the square of the frequency. In sensor or audio applications, with relatively low signal bandwidth, flicker noise may also be relevant with a dependence on inverse cube of the frequency. Phase noise can be simulated very efficiently with the periodic state circuit simulators (PSS) built in commercial design software.

The Input-Referred Phase Noise (IRPN) can easily be calculated if the phase noise spectral density plot is known (Fig. 2.7c) using the definition of phase noise and the gain constant of the VCO, K_{VCO}. The IRPN is simply the voltage source that we need to put at the input of an ideal noiseless VCO to obtain the same phase noise as in the real VCO. Figure 2.7a depicts a system-level model of a VCO where the phase noise $\varphi(t)$ is shown as an additive component to the VCO phase signal. Figure 2.7b shows an input voltage source $r(t)$ representing the IRPN that needs to be added to an ideal noiseless VCO. The mathematical relationship between $r(t)$ and $\Phi(t)$ is a scaling factor and a derivative that accounts for the phase being the integral of the frequency. A detailed mathematical explanation can be found in [Car18a], [Car18b].

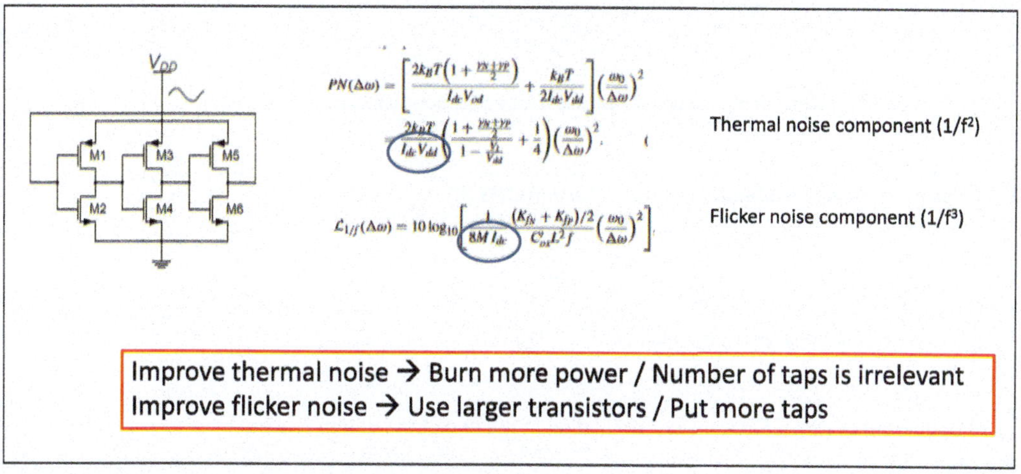

Fig. 2.8 Phase noise in a ring oscillator with M taps

Ring oscillators have been analyzed extensively in the past to establish a model that permits to theoretically predict their phase noise. To name a few, the following papers survey the topic: [Raz19], [Kim10], [Raz96], [Pep19]. Although the results of each analysis differ in small details, there is a common agreement on the dependence of phase noise on the transistor sizes, the number of taps, and the power consumption. The behavior of the oscillator differs for thermal noise and for flicker noise. For these reasons they are typically analyzed separately. The thermal noise component depends inversely on the power consumption of the VCO, as could be expected, but in most cases, is independent on the number of taps [Raz19], [Raz96], [Pep19]. On the other hand, flicker noise behavior depends on the current consumed and on the number of taps (M). These components result into thermal noise and flicker noise with flat and $1/f$ spectral noise densities when transformed into IRPN. To improve the thermal component of the IRPN of a ring-oscillator VCO, we have to increase the power consumed by the VCO if we keep the oscillation frequency to comply with the desired quantization noise. To improve the flicker component we can increase the number of taps, but as explained before, this also represents an increase in the digital logic power and complexity (Fig. 2.8).

1) Calculate $S_\phi(f)$ (may be simulated with a periodic steady state noise simulator, PNOISE)

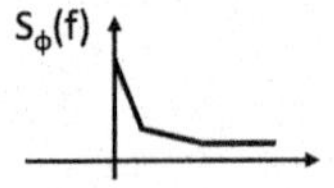

2) Calculate the input-referred spectral noise density $S_r(f)$

$$S_r(\Delta f) = S_\phi(\Delta f)\left(\frac{2\pi\Delta f}{2\pi k_d f_0}\right)^2$$

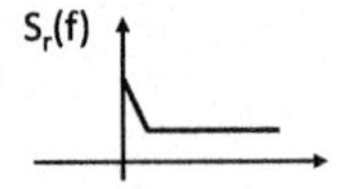

3) Integrate over input signal bandwidth

$$SNR = 10 \cdot log_{10}\left(\frac{P_{signal}}{P_{noise}}\right) = 10 \cdot log_{10}\left(\frac{v_{in\text{-}peak}^2/2}{\int_{f_{Lo}}^{f_{Hi}} S_r(\Delta f)d\Delta f}\right)$$

4) Double check with a full transient simulation at the end of the design process

Fig. 2.9 Method to (quickly) predict the SNR of a VCO-ADC

Here we summarize a design methodology that enables us to dimension the oscillator of a VCO-ADC once the quantization noise has set the effective oscillation frequency.

A first step is to simulate the spectral density of the phase noise using a commercial periodic steady-state noise (PNOISE) simulator. Note that when configuring the simulation we must define a minimum frequency for our analysis, which cannot be zero. The maximum frequency should at least extend till the end of the input analog bandwidth of our ADC (see step 1 in Fig. 2.9).

With the phase noise spectral density we apply the equations from [Car18b] (reproduced in Fig. 2.9) to calculate the IRPN spectral density, which should look like a 1/*f* part reflecting flicker noise contribution and a white noise part representing the thermal noise contribution (see step 2 in Fig. 2.9).

Finally, to calculate the predicted SNR of the ADC due to the thermal and flicker noise, we integrate the IRPN spectral density over the required signal bandwidth to find the noise power. We obviously assume here that quantization noise is not dominating the SNR and therefore SQNR > SNR. The signal power can be calculated by directly using the full-scale input voltage (step 3 in Fig. 2.9).

This process can be quite fast using conventional CAD tools and allows us to interactively optimize the transistor sizes. To clear out any modeling and calculation inaccuracy, we perform a final transient noise simulation with the definitive circuit, once the oscillator has been dimensioned (step 4 in Fig. 2.9).

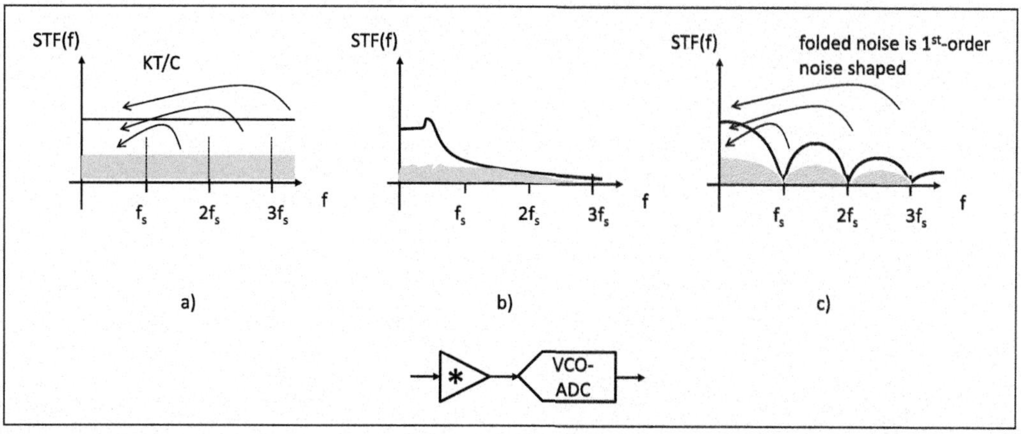

Fig. 2.10 Do VCO-ADCs have noise folding? (**a**) Switched-capacitor ΣΔ, (**b**) continuous-time ΣΔ, (**c**) open-loop VCO-ADC

A well-known phenomenon of sampling ADCs is noise folding. Noise folding happens whenever we sample a signal with a sample and hold. It dictates the required size of the sampling capacitor C through the well-known integrated noise power estimation formula kT/C. This is the typical situation in SAR ADCs using a capacitive array or in switched-capacitor Sigma-Delta ADCs (see Fig. 2.10a).

In continuous-time Sigma-Delta modulators, the situation is very different because the sampling takes place at the quantizer and noise is typically dominated by the contribution of the first integrator. Input-referred noise in continuous-time Sigma-Delta modulators is filtered by the Signal Transfer Function (STF) of the modulator which is a low-pass filter. Therefore, continuous-time Sigma-Delta modulators can be assumed to be free from noise folding effects (see Fig. 2.10b).

What can be said about noise folding in VCO-ADCs? This can simply be answered by understanding that VCO-ADCs also have a STF that is a sinc function (see Fig. 1.18). In this case, the input-referred thermal noise may have a spectral density reaching up to high frequencies, but it is filtered by the sinc transfer function before being folded to the Nyquist band (see Fig. 2.10c). Therefore, we can say that VCO-ADCs are affected by noise folding, but noise folded into the signal band is first-order noise shaped just like the quantization noise.

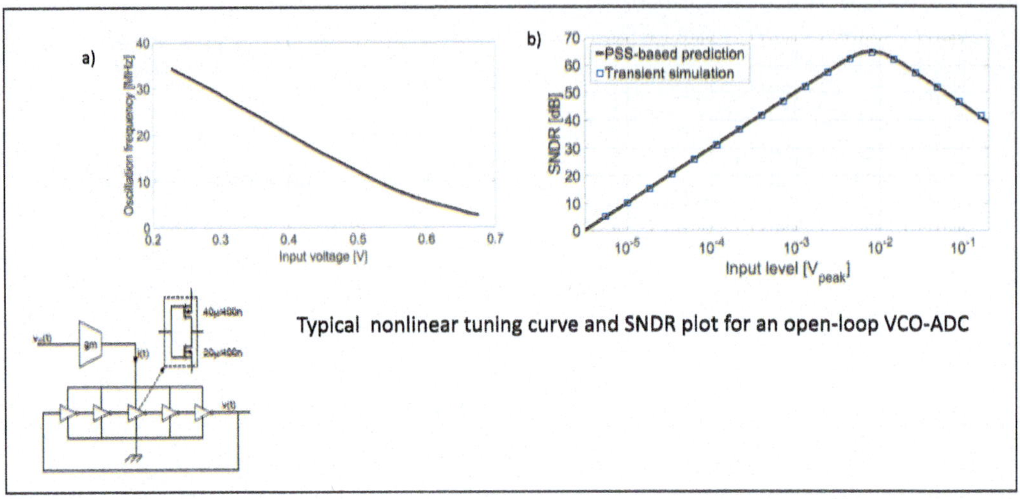

Fig. 2.11 VCO distortion

Even when using the pseudo-differential architecture of Fig. 1.8, the nonlinearity of the VCO tuning curve and the harmonic distortion that this causes is the main limitation for the widespread use of VCO-ADCs. Most of novel VCO-ADC architectures to date try to overcome this problem. In the case of ring oscillators, there is a minimum supply voltage to keep the ring oscillating and also, as V_{dd} increases, the slope of the voltage to frequency tuning function curbs down. The influence of a non-linear oscillator tuning curve can easily be estimated using microelectronic CAD tools, like in the case of phase noise.

Oscillator analysis based on periodic steady-state simulation permits to plot the tuning curve of the oscillator (see Fig. 2.11a). Then, a polynomial fit of the curve to a moderate-degree polynomial, allows to calculate the distortion coefficients for the dominant harmonics. These distortion coefficients together with the IRPN calculated before allow to determine the dynamic range of a VCO-ADC with very good accuracy (see [Car18b] for details) without resorting to very lengthy time-domain simulations (see Fig. 2.11b).

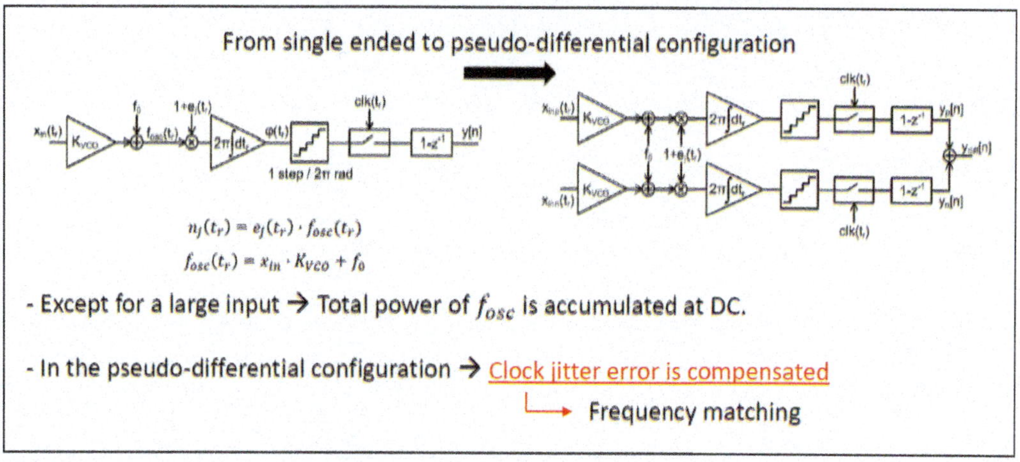

Fig. 2.12 Clock jitter in VCO-ADCs

Clock jitter is of concern in continuous-time Sigma-Delta modulators due to the matching required by feedback DACs and digital output data. Sampled ADCs like switched-capacitor Sigma-Delta modulators and SAR ADCs are less sensitive to jitter as only the sampling-induced jitter is relevant. In the case of VCO-ADCs we can say that its sensitivity to jitter is lower than that of continuous-time Sigma-Delta modulators. The jitter behavior of VCO-ADCs has thoroughly been studied in [Car19]. The pseudo-differential architecture of Fig. 1.8 in particular is resilient to clock jitter if the rest frequency of the two VCOs is carefully matched. Under this matching condition, clock jitter errors, shown in Fig. 2.12 as additive components, become a common-mode signal, which is canceled by the differential structure (see a detailed explanation in [Car19]).

Chapter 3
Overview of VCO-ADC Architectures

G. Gielen et al., *Time-encoding VCO-ADCs for Integrated Systems-on-Chip*,
https://doi.org/10.1007/978-3-030-88067-5_3

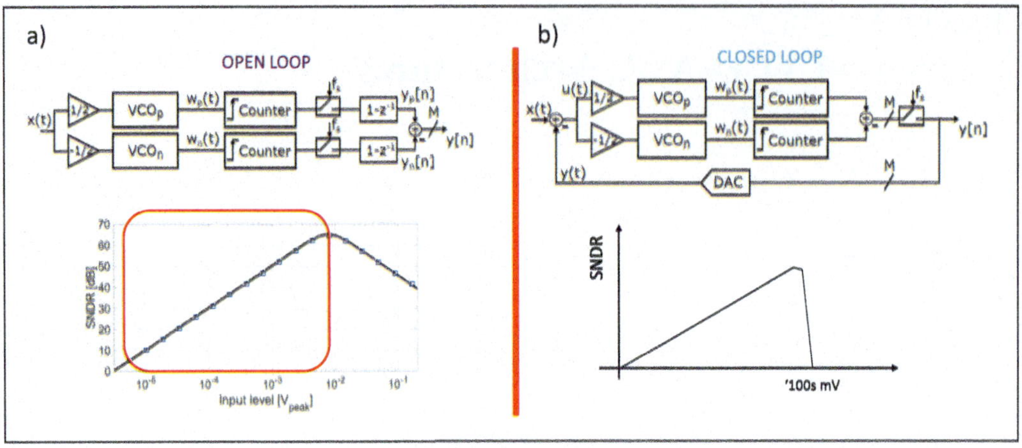

Fig. 3.1 Open-loop versus closed-loop architectures

Until now, we have discussed the main system-and-circuit related topics about the reference VCO-ADC architecture depicted back in Fig. 1.7. However, practical VCO-ADCs can adopt many other architectures to improve distortion, dynamic range, or power consumption based on the VCO-ADC of Fig. 1.7. We are especially interested in ADC implementations that use mostly digital circuits, which can be enabling architectures for the mixed-signal SoCs displayed in Fig. 1.1. There are two main kinds of such architectures, open-loop architectures (see Fig. 3.1a), and closed-loop architectures (see Fig. 3.1b). We are going to briefly introduce their properties.

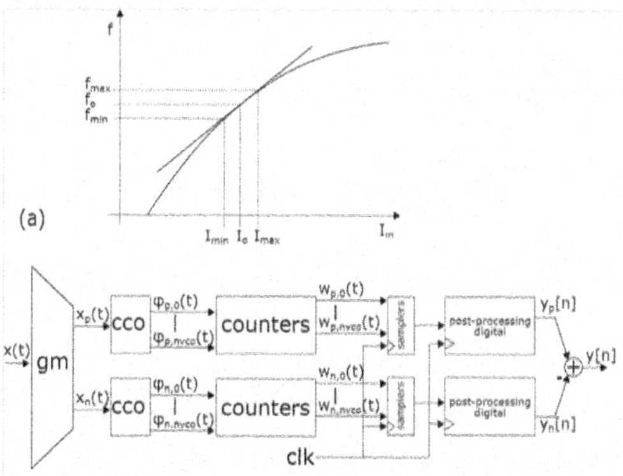

Advantages:
1) Simple !!
2) Input range is small...no preamps needed
3) No need to use a DAC with matching
4) Input can be directly connected to high-impedance source (capacitive MEMS, biopotential, etc)
5) Little clock jitter influence (if VCOs are matched)

Disadvantages:
1) Linearity of VCO is not compensated
2) f_0 may need to be very large to overcome small input signal range

Fig. 3.2 First-order open-loop VCO-ADCs

Open-loop architectures correspond to the structure of Fig. 1.7, i.e., a pseudo-differential circuit with an analog preamplifier (a transconductor or source follower) that drives two twin VCOs implemented with ring oscillators. Then, a digital decoding circuit is used to sample the oscillation and to generate a digital output code. If the purpose of the ADC is to convert a wide-bandwidth signal, like in a communications application, we may use the standard XOR—flip-flop decoder [Kim10]. If we are interested in a highly sensitive, high-input impedance ADC, like in a sensor application, we may use a counting architecture [Qui20]. The main advantage of open-loop VCO-ADCs is their simplicity. There is no need to design and implement a high-linearity DAC with many quantization levels or wide-bandwidth opamps, like in a continuous-time Sigma-Delta modulator. The circuit only requires a simple analog block (the transconductor or source follower) and the rest are digital blocks that can be designed even with a digital synthesis tool. Also, power is optimized because the biasing current of the analog preamplifier can be reused to bias the VCOs. The jitter behavior can be optimized as described in Fig. 2.12. As the main disadvantage of open-loop VCO-ADCs, the nonlinearity of the VCO tuning curve manifests itself at the output of the ADC, causing distortion and degrading the SNDR (Fig. 3.2).

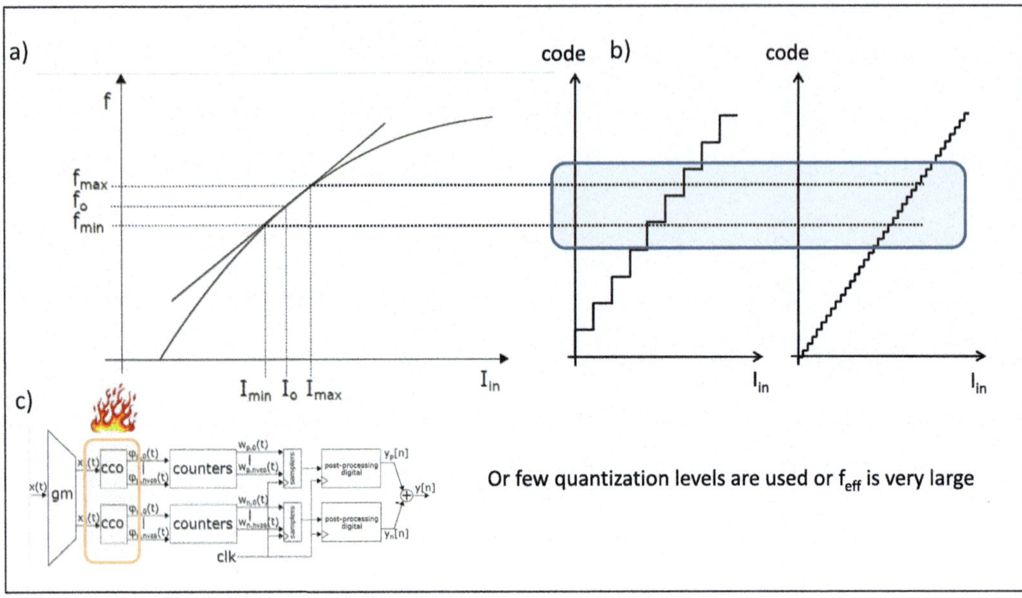

Fig. 3.3 Problem of first-order open-loop VCO-ADCs

There are two ways to improve the linearity of open-loop ADCs. A first method is to use calibration. Calibration can be implemented digitally with look-up tables or polynomial functions after the decoding logic. Even simple, power-and area-efficient digital calibration circuits can be very effective in improving the linearity for ENOB values between 10 and 12 bits [Wu19]. The main problems of calibration are the calculation of calibration coefficients and the robustness of the calibration against PVT variations.

The second way is to live with nonlinearity as a minor problem for the sake of simplicity. Let us assume that our VCO has the tuning curve of Fig. 3.3a. We can reasonably assume that this curve is sufficiently linear if only a small input portion is used (between f_{min} and f_{max} in Fig. 3.3a). With the methodology of Fig. 2.5 we would calculate an effective frequency that would require to use the whole tuning curve (see Fig. 3.3b). However, we can raise the effective frequency to a higher-than-required value and the same resolution will be achieved but with a much smaller voltage variation (see Fig. 3.3c). Then, the linearity of the ADC will be improved at the expense of forcing the oscillator to oscillate very fast.

This approach is somehow suboptimal in power but may be especially interesting because the analog circuitry is minimum, which favors integration in nanometer CMOS. There are also applications such as audio [Qui20] or biomedical ADCs [Jia17b] where the distortion of large signals is of lesser importance in exchange for a high dynamic range. The power efficiency of this kind of ADCs can be very good if we need to connect a high-output-impedance source (like a capacitive sensor or a biomedical probe) since the power of the coupling amplifier is shared by the VCOs.

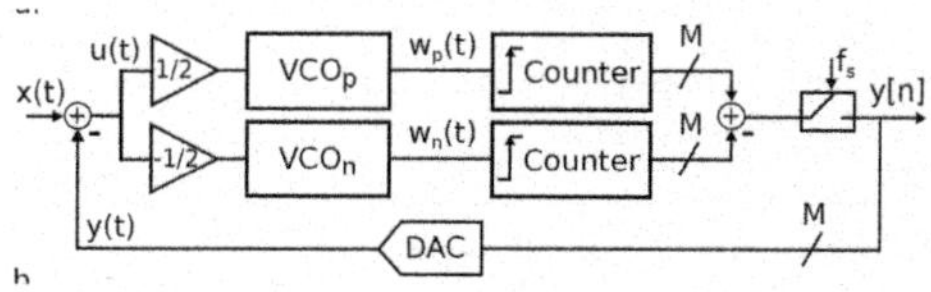

Advantages:
1) VCO nonlinearity is attenuated by K_{VCO}
2) Input range can be Vdd
3) Still no opamps are needed

Disadvantages:
1) Needs a DAC. Not all configurations do DWA, most do CLA
2) A preamplifier is needed to connect a high-impedance source (FoM misconception!!)
3) Depending on the DAC Architecture, clock jitter may be relevant

Fig. 3.4 First-order closed-loop VCO-ADCs

VCO-ADCs can also be implemented as a closed-loop system without requiring a complicated analog hardware. By using the fact that a ring oscillator can be seen as an oscillator plus an integrating function (counter), we can reproduce the operation of a conventional continuous-time Sigma-Delta modulator (see Fig. 3.4). In this case, a pseudo-differential circuit can also be used, but the ring oscillator outputs are processed in a different manner [Lee15]. A feedback DAC closes the loop from the output back to the input, such that the input of the VCOs become a virtual ground node and the large gain of the VCO helps to linearize the tuning curve.

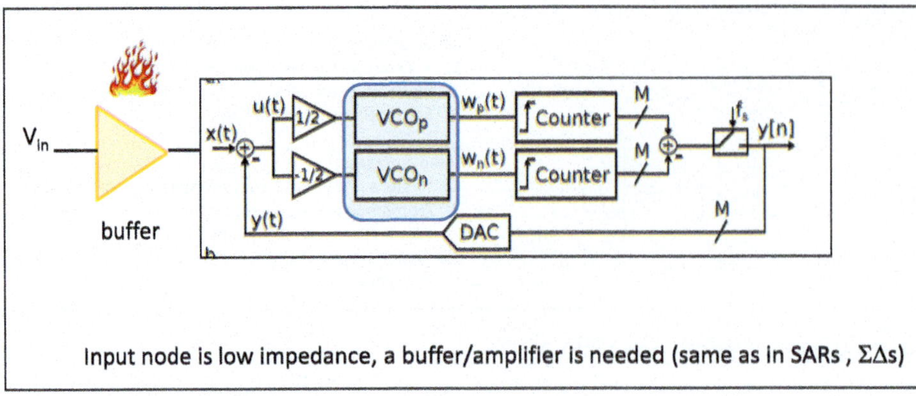

Fig. 3.5 Problem of first-order closed-loop VCO-ADCs

In closed-loop VCO-ADCs, the virtual ground input represents a low-impedance point. Therefore, the input voltage to be acquired needs to be converted into a current by means of a series resistor. Hence, a voltage buffer is required if a high-impedance source is to be used. Therefore, closed-loop architectures are apparently optimal in power consumption as the VCO runs at the minimum effective frequency (see Fig. 2.3) which leads to an excellent Figure of Merit (FoM). However, if we require a high-input impedance and we include the power consumption of the input buffer, the overall total power can even exceed that of open-loop architectures (Fig. 3.5).

Part II

VCO-ADC Circuits and Systems

Chapter 4
Ring Oscillators and Their Design Methodology

G. Gielen et al., *Time-encoding VCO-ADCs for Integrated Systems-on-Chip*,
https://doi.org/10.1007/978-3-030-88067-5_4

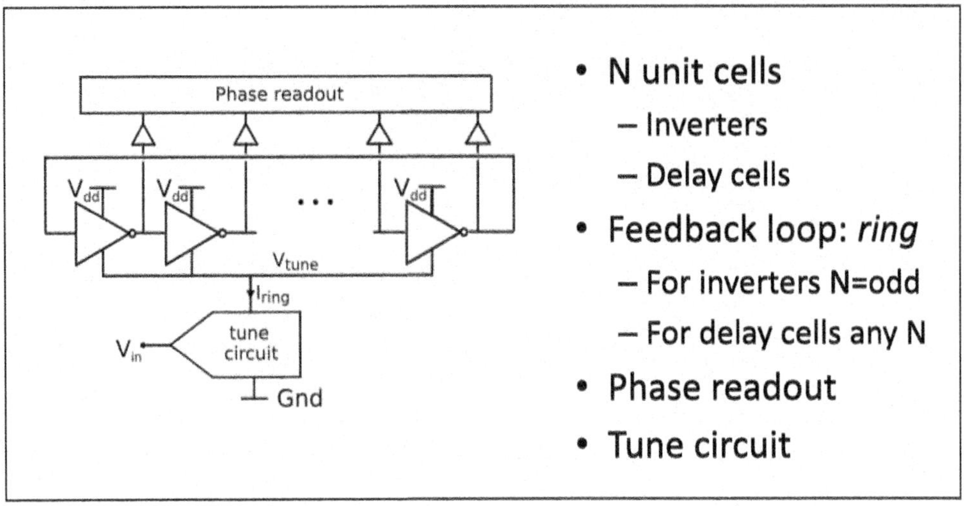

Fig. 4.1 Ring oscillator circuit

Figure 4.1 shows a conceptual ring oscillator VCO-ADC circuit. We see a number of unit cells, N in total. In Fig. 4.1, these unit cells are inverters, but more general configurations where the unit cells are delay elements are also possible. These unit cells are configured in a feedback loop to form a ring oscillator (see Fig. 2.2). We want the ring to oscillate and if the unit elements are inverters, this requires that the number of unit elements is odd. Otherwise we would obtain a stable latched state which is not what we want here.

If the unit elements are delay elements, we need a way to obtain a signal inversion somewhere in the loop, because otherwise we will also end up in an undesired stable latched state. We will see below that most practical circuits for delay elements are fully differential. Hence this signal inversion can readily be obtained by cross-coupling the differential signal wires. With this in place, we can also use an even number N of elements in a ring with delay elements instead of inverters. We will see below that the delay elements can also be formed with "nearly digital" circuits. Remember, the inverters in practice are 2-transistor CMOS inverters.

The output of every unit cell is connected to a phase readout circuit (see Fig. 1.13). And finally there is some tune circuit that couples the overall VCO input to the actual ring.

We will now for the moment put the tune circuit and the phase readout aside and focus on the actual ring first.

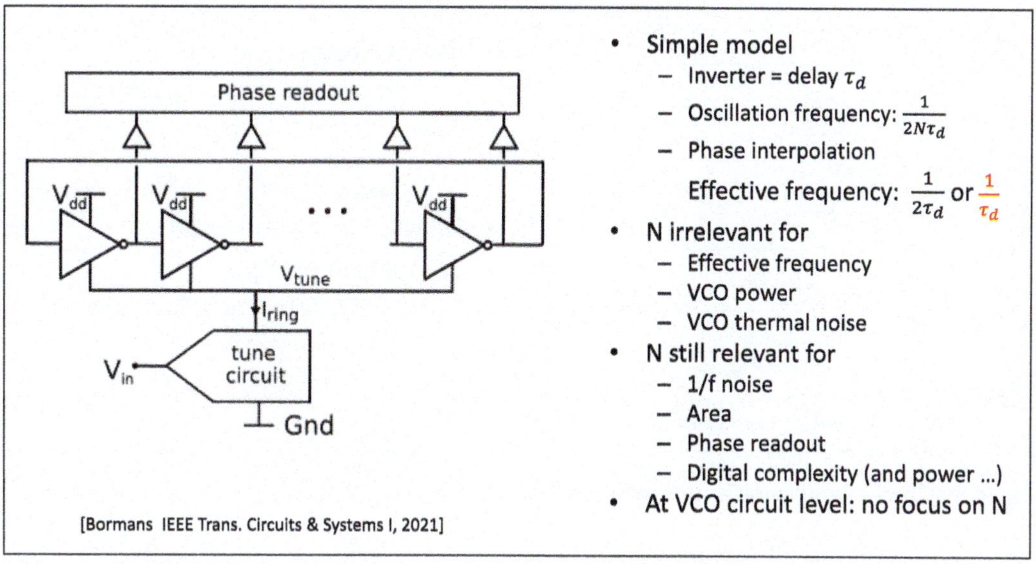

Fig. 4.2 Effective frequency

The oscillation frequency of this circuit was already analyzed above (Figs. 2.3 and 2.4) and is based on the simplest possible model for the unit cells (Fig. 4.2). I.e., it operates full "digital" in the sense that the output is always either "high" or "low" and the dynamic behavior is modelled as a delay τ_d. It was shown that the corresponding oscillation frequency is $1/(2N\tau_d)$. But if we read all phases, we actually detect the rising edges at an effective frequency $f_{eff} = 1/(2\tau_d)$. If also the falling edges are detected, this increases the effective oscillation frequency by a factor of 2, and hence the effective frequency becomes $f_{eff} = 1/\tau_d$. Note that this happens in the widely used XOR-based readout described on Fig. 2.3. This effective frequency is what sets the quantization noise performance and is what should be used in conjunction with the SQNR equation derived on Fig. 1.12.

From the expression above it is clear that the effective frequency no longer depends on N, the number of unit cells. Now, it is important to realize that most "analog" performance metrics are also independent of N. First, the VCO power is independent on N. This can be understood by observing that such a CMOS inverter only consumes dynamic power (i.e., it only consumes power while it is switching). As there is only 1 inverter switching at the same time, regardless of the number of cells in the ring, the power consumption is independent of N. A similar argument holds for the VCO input-referred white (thermal) noise. Indeed, there is only one cell at the time that processes an edge. And in our system only the edges matter. This way there is always only one cell that can contribute to jitter on the edge position and again this independent from N. While these are only qualitative arguments, a more detailed study [Bor21] fully confirms this.

There is one important "analog" property that does depend on N: the flicker ($1/f$) noise. This was already known a long time for the output phase noise [Abi06] and was extended for the input-referred noise in [Bor21] where it was shown that the input-referred $1/f$ noise is proportional to $1/\sqrt{N}$. This implies that the $1/f$ noise performance can always be improved by increasing N.

Note that according to [Abi06], some classes of analog ring oscillators have a different behavior, but these are out of the scope of this discussion.

While the raw "analog" properties of the ring are mostly independent on N, the opposite is true for many other (system) level metrics. Clearly the area is simply proportional to N. Also the complexity and the power consumption of the digital and phase readout circuits very heavily depend on N. Thanks to the fact that the "Analog" performance is largely independent on N, in practice the decision on N will be set by these considerations.

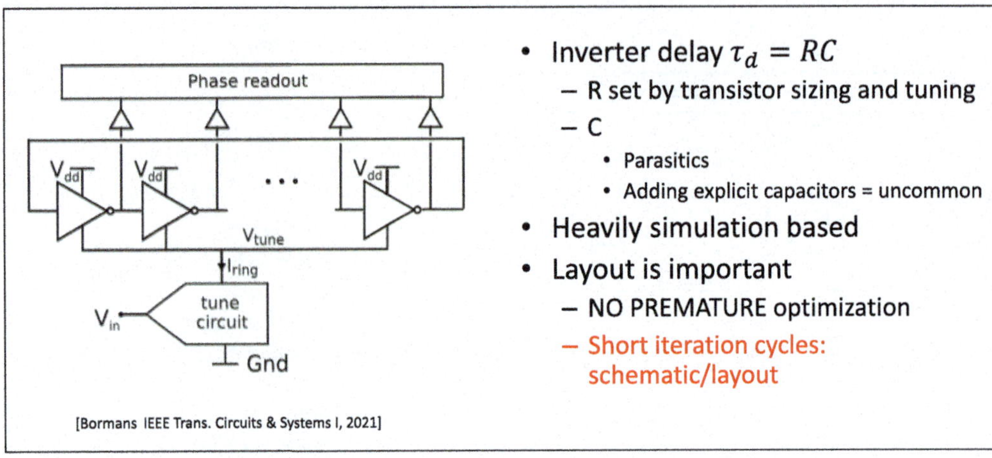

Fig. 4.3 Case study: inverter unit cell

We will now elaborate the circuit-level design of the elements in the ring. From the above it is clear that the most important parameter is the inverter delay τ_d. In general we can always state that this delay will be of the form $\tau_d = RC$, where R is some impedance parameter of the inverter and is related to its drive strength. The capacitor C is the load seen at the output of each inverter. In most cases this load will be dominated by circuit parasitics. In particular the gate-source and gate-drain parasitic capacitors will form an important contribution. It is thinkable to also add explicit capacitors (which can be done without a power or thermal noise penalty, see the detailed study of [Bor21]). However, this is quite uncommon because it lowers the effective oscillation frequency and hence leads to a reduced SQNR (Fig. 4.3).

In some cases, lowering the effective oscillation frequency is not a big deal or could even be desired (e.g., to simplify the readout). This situation occurs, for example, in low-bandwidth systems where even for a modest effective VCO frequency the oversampling ratio would already be more than high enough. But even in this case it is better to increase the transistor length L. This will also lower the effective oscillation frequency, but at the same time it will improve the low-frequency $1/f$ noise, see [Bor21], and hence gives a better overall performance.

While rough estimations of the parasitic capacitors for back-of-an-envelope calculations are adequate to explore trends and trade-offs, the actual design requires extensive simulations. This is even more important in scaled technologies with small feature size. In such an ultra-deep submicron technology in most cases the interconnection parasitics will also contribute significantly to the inverter delay. Simulation results from an extracted layout can easily deviate a factor 2 from the raw schematic. It is important to realize this up front and to adjust the design strategy with this in mind. An important consequence is that circuit optimization relying only on a schematic (without layout) makes no sense at all. Instead, a first layout should be generated as fast as possible in the design process and also subsequent design iterations should be kept as short as possible and should include the layout.

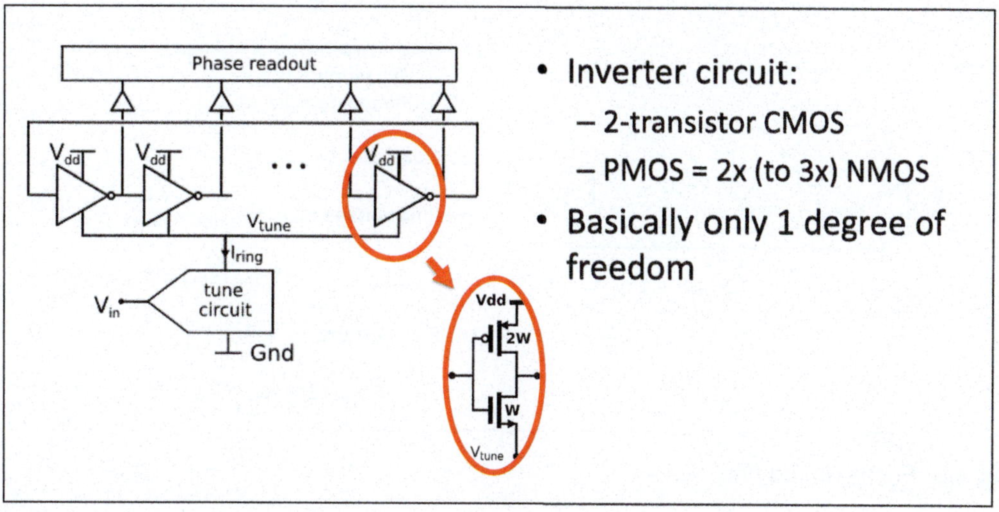

Fig. 4.4 Inverter unit cell: what is there to design?

Let us now focus on the circuit design of the ring. Since we have established above that the decision on the number of cells in the ring (*N*) should be made at the system level, the only thing that remains to be sized are the transistors in the inverters. These inverters are standard 2-transistor CMOS inverters, where usually the PMOS transistor is sized with an aspect ratio (*W*/*L*) of around 2–3 times that of the NMOS transistor. This is to ensure that the circuit has a similar rise and fall time. Within our thinking on this circuit this also means that the rising-edge delay matches the falling-edge delay which is desired (Fig. 4.4).

As a result, the circuit design simply boils down to sizing the overall inverter drive strength (set by the *W*/*L* of the NMOS transistor).

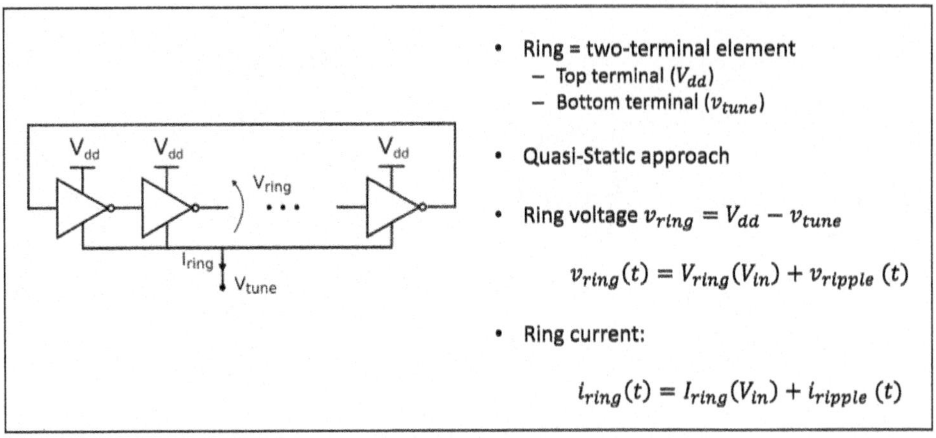

Fig. 4.5 Focus on the ring now

Let us now strip the ring out of the overall ADC and take a bird's eye view. From this perspective, we can consider the inverters and the connections forming the actual ring as the circuit's internals. Overall there are two "external" terminals: a "top terminal," in this case connected to the supply voltage V_{dd}, and a "bottom terminal," connected to the tune voltage v_{tune}. In this way we can think of the ring as a two-terminal element that is defined by the voltage v_{ring} over the two terminals and the current i_{ring} flowing in these terminals (Fig. 4.5).

Here, we use a lowercase letter to explicitly indicate that these signals are time dependent. We will use uppercase letters to denote signals that are not time dependent. To study the relationship between the voltage v_{ring} and the current i_{ring}, we will follow a quasi-static approach. We will assume that both signals consist of a quasi-static component (V_{ring} and I_{ring}, respectively) and a high-frequency ripple (v_{ripple} and i_{ripple}, respectively). The quasi-static component depends on the biasing of the ring. It also depends on the overall VCO input voltage V_{in}. Hence this component is not really constant, but it is at a much lower frequency than the ripple component (as will be illustrated below). Hence its rate of change is negligible compared to the ripple components. For this reason we can consider this signal as quasi-static.

The ripple component is associated with the oscillation. As we will see below it occurs at half the effective VCO frequency. In any VCO-ADC this frequency is much higher than the usable input signal frequency.

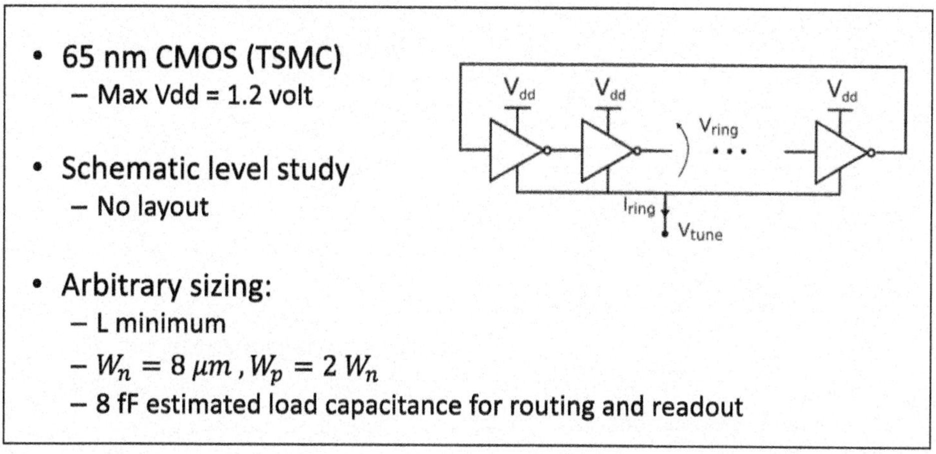

Fig. 4.6 Example case study

To illustrate the discussion elaborated below, we will use an example case study. For this we use a commercially available 65 nm CMOS technology with a maximum supply voltage of 1.2 V. Since our goal is to illustrate the trends, we will restrict ourselves to results at the schematic level (neglecting layout effects) (Fig. 4.6).

Remember that in an actual design, you must take layout effects as early as possible into account!

To illustrate the discussion we will use an arbitrary sizing. We will later explain how the sizing should be adjusted to take different specifications into account.

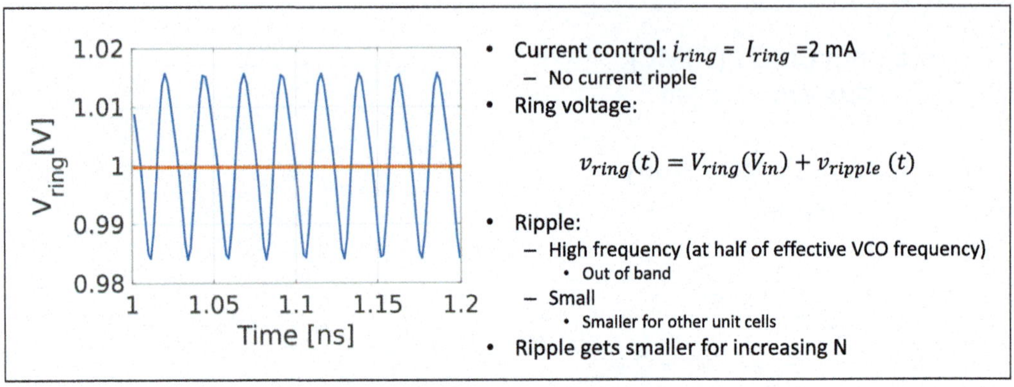

Fig. 4.7 VCO waveforms: current control (inverter unit cell)

Let us now investigate the ring voltages and currents for our case study. First we consider the case of current control, i.e., we drive the ring with a fixed current. This means that the current ripple is zero, but the voltage ripple will not be zero. Figure 4.7 shows a simulation result for the ring voltage waveform for the case where the ring was driven with an arbitrarily chosen current of 2 mA. We can see that there is a large quasi-static component of $V_{\text{ring}} = 1$ V in the ring voltage waveform. As expected, there is also a ripple waveform which is associated with the high-frequency VCO oscillation. A first important observation is that the ripple is at half the **effective** oscillation frequency f_{eff} and not at the actual oscillation frequency. Remember that this effective oscillation frequency is independent from N, the number of unit cells in the ring. Normally this effective oscillation frequency is very far out of the useful ADC signal band. A second important observation is that the ripple is quite small: around 1.5% of the quasi-static component. We will later discuss other types of unit cells for which the ripple will be even smaller.

The represented waveform figure was initially obtained for a ring with only $N = 9$ elements. Afterwards, the simulation was repeated for $N = 17, 33, 65, 129$ elements. It was found that the ripple decreases for higher values of N. This makes sense because the ripple is at a high frequency and is smoothened by (parasitic) capacitance at the tune node. When increasing N, the number of transistors connected to the tune node increases as well and hence the capacitance at this node increases, leading to this effect.

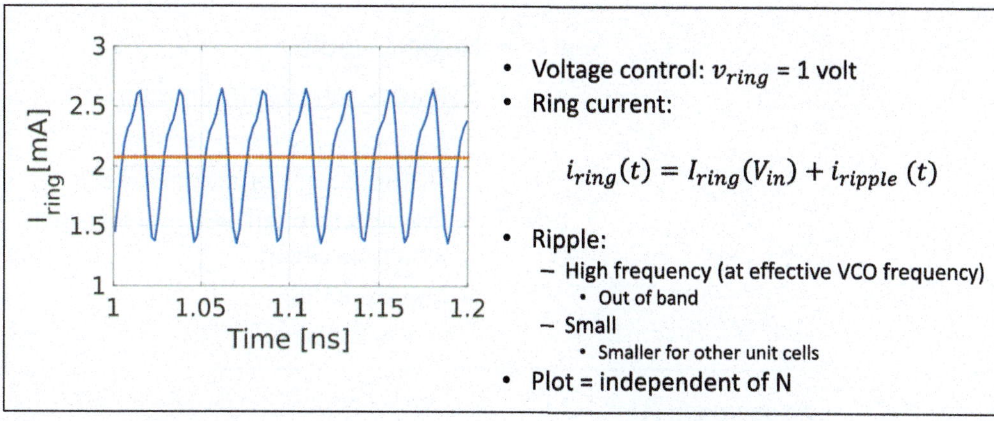

Fig. 4.8 VCO Waveforms: voltage control (inverter unit cell)

Next, we consider the case of voltage control, i.e., we apply a fixed voltage to the ring. Hence the voltage ripple is zero, but the current ripple will not be zero. Figure 4.8 shows a simulation result for the ring current waveform for the case where a voltage of 1 V is applied to the ring. Now we see a relatively large quasi-static component of $I_{\text{ring}} = 2$ mA in the ring current waveform. As expected, there is also a ripple waveform which is associated with the high-frequency VCO oscillation.

With regard to this current ripple waveform, we can make very similar observations as for the voltage ripple waveform on the previous figure. It is at half the effective oscillation frequency f_{eff} and not at the actual oscillation frequency. Also, the magnitude can be considered to be relatively small, but in this case it amounts to 25% of the quasi-static component. For the other types of delay cells (discussed below), the ripple is much smaller (a few percent of the quasi-static component).

Unlike for the case of current control, it is found now that the represented waveform is identical for different values of N, the number of elements in the ring (with simulations for $N = 9, 17, 33, 65, 129$).

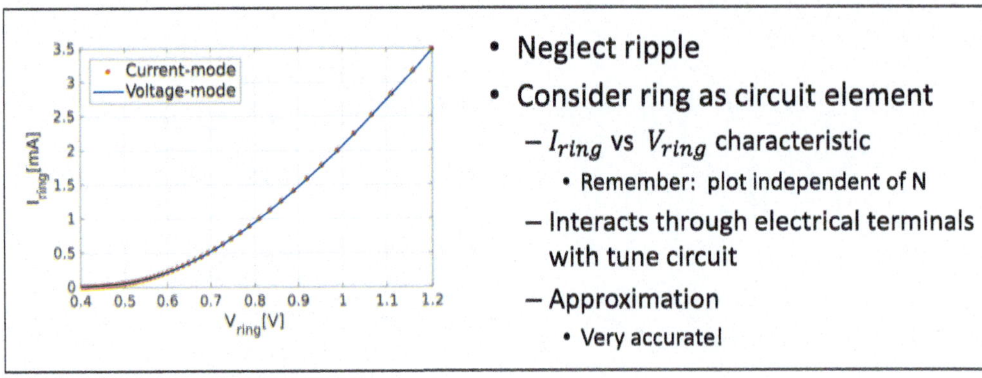

Fig. 4.9 Quasi-static approach (1)

These observations allow us to move to the next step where we will ignore the ripple and focus only on the quasi-static components I_{ring} and V_{ring}. Now, we can investigate the relationship between these quasi-static components. For this, we have done two simulation batches. First, we have driven the ring in voltage control for a fixed ring voltage V_{ring}, and we have captured the corresponding quasi-static current I_{ring} (by stripping off the current ripple from the current waveform). This experiment has been repeated for varying values of the V_{ring} voltage. The corresponding results are represented in Fig. 4.9 as the blue (continuous) curve. Then in a second batch of simulations, we have driven the ring in current control with a fixed current I_{ring} and we have captured the corresponding quasi-static component of the ring voltage V_{ring} (by stripping off the ripple from the waveform). Again, this experiment was repeated for varying values of I_{ring}. The corresponding results are represented in Fig. 4.9 as the red (dotted) curve.

When observing the curve, it is clear that the curves are nearly identical. When studying the detailed simulation data, it can be found that the curves are not exactly identical, but the differences are always less than 0.1% and hence we can state that they are negligible. From this, we can conclude that for what concerns the quasi-static voltage and current components, the ripple can be neglected. Note that this is not an obvious result as the ring circuit (as can be seen on Fig. 4.9) clearly is a nonlinear circuit and hence waveform superposition normally is not valid.

With this in mind, we can now look at the ring as a two-terminal circuit, defined by the represented I_{ring} vs. V_{ring} characteristic. The interaction with the rest of the connected circuits (in particular the tune circuit) then occurs entirely through these two terminals and obeys this I_{ring} vs. V_{ring} characteristic. While strictly speaking this is an approximation, from the plot it is clear that it is a very accurate approximation.

Finally, it should be kept in mind that the I_{ring} vs. V_{ring} characteristic only depends on the invertor sizing and very importantly, not on N, the number of elements in the ring. It also does not depend on the load capacitance C_{load} on the internal nodes of the ring.

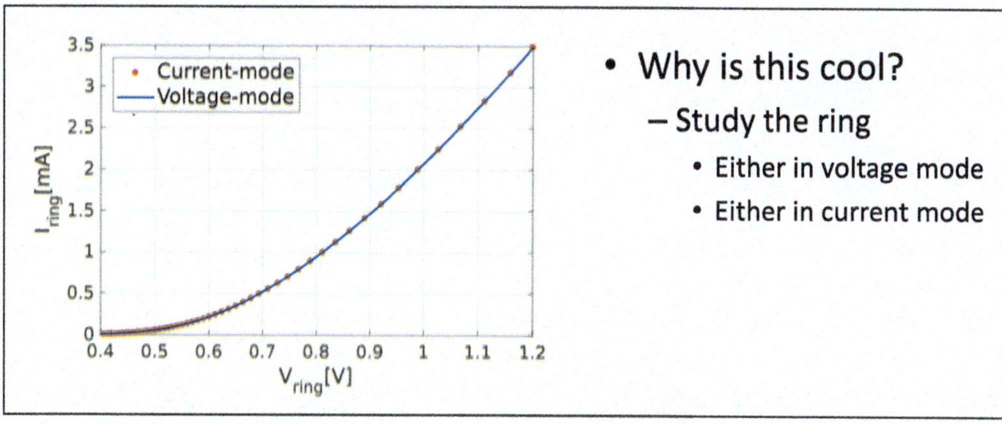

Fig. 4.10 Quasi-static approach (2)

Now we can ask ourselves: for what is this helpful?

First, it establishes a clear relationship between the (quasi-static component of the) current in and the voltage across the ring. This means that all other ring characteristics can be studied either in current mode or in voltage mode, and the other characteristics can be obtained through this *I*/*V* characteristic. For example, if you have captured the frequency vs. voltage characteristic (i.e., a voltage-mode characteristic), it can easily be converted to a frequency vs. current characteristic by mapping the voltages to the corresponding currents through this *I*/*V* characteristic. A similar argument holds for the current-mode approach (Fig. 4.10).

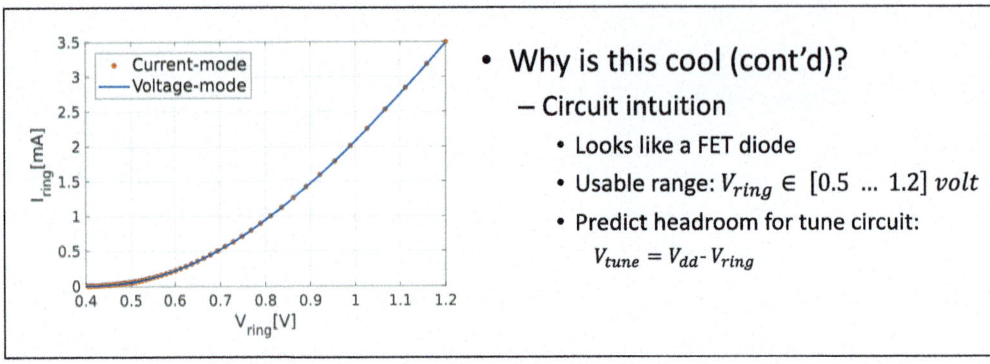

Fig. 4.11 Quasi-static approach (3)

A second advantage of the approach is that it gives circuit intuition. Most readers will notice the remarkable resemblance of this *I*/*V* characteristic with that of a diode-connected FET. This means that from an electrical point of view all circuit intuition that designers have on this ubiquitously known device can be transferred to the electrical behavior of the ring. For example, we can immediately see the usable range of the ring: we need at least 0.5 V over the ring. In the technology that we used as an example here, the maximum supply voltage is 1.2 V. Hence we can immediately predict the headroom that is available for the tune circuit, which in this technology is always less than 0.7 V (Fig. 4.11).

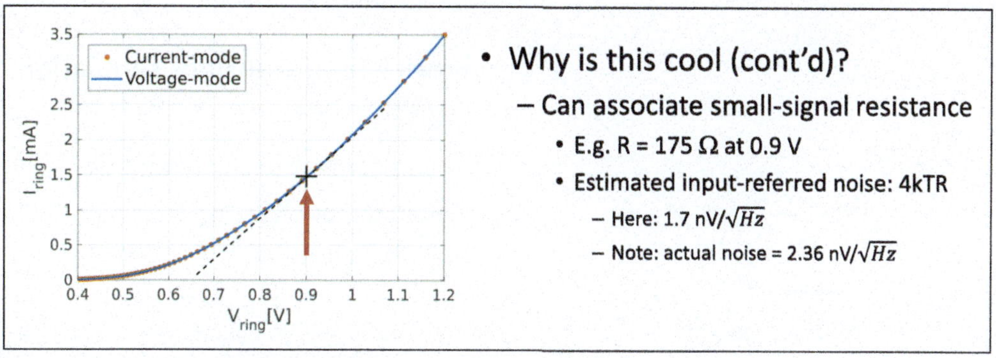

Fig. 4.12 Quasi-static approach (4)

Next to the intuition on the large-signal behavior of the ring, we can also obtain intuition on the small-signal electrical behavior. Consider, for example, a ring biased around 0.9 V. Then we can find the small-signal resistance of the ring. For the example shown here, the small-signal resistance is around 175 Ω. This allows to get a quick assessment of the expected electrical white voltage noise spectral density as 4kTR. For a resistor with a resistance of 175 Ω this would be 1.7 nV/$\sqrt{\text{Hz}}$ (Fig. 4.12).

For this case, the actual overall input-referred phase is 2.36 nV/$\sqrt{\text{Hz}}$ (see discussion later on).

We do not expect the quick assessment to be *very* accurate, because what interests us is the output phase noise referred to the input and this could be affected by many other factors that are not included in the electrical input behavior. But as we can see, the quick assessment gives the correct order of magnitude.

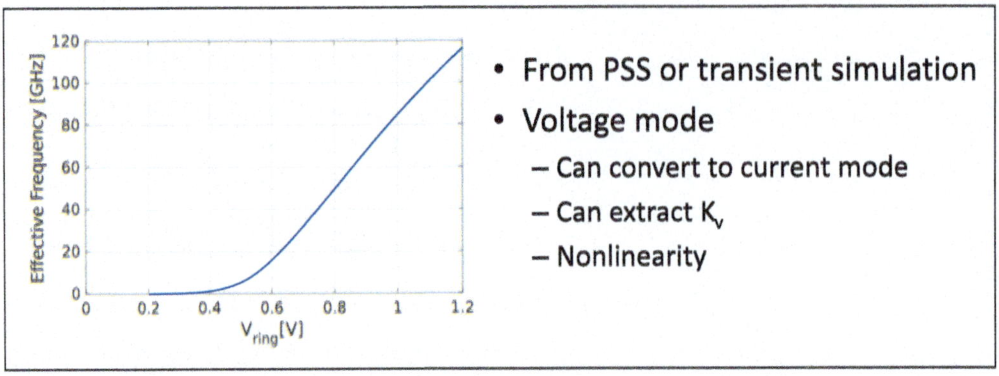

Fig. 4.13 Ring characteristic: frequency

After having studied the electrical behavior of the ring, we are of course interested in the frequency characteristic of the ring. A transistor-level simulation result (schematic level) of our case study is shown above. Here the ring was driven by a fixed value of the ring voltage V_{ring} and the corresponding value of the oscillation frequency was captured. This was repeated for different values of the ring voltage V_{ring} to obtain the curve. In principle the oscillation frequency can be obtained through a transient simulation, but in our experience it is much more efficient to use a Periodic Steady-State (PSS) simulation as available in commercial simulators (Fig. 4.13).

The represented curve corresponds to voltage mode, but as explained above, by using the ring's *I*/*V* characteristic it can immediately be converted to a current-mode frequency vs. current characteristic. Furthermore, this plot allows to study the important VCO parameters such as the VCO sensitivity K_V and the VCO (non)linearity.

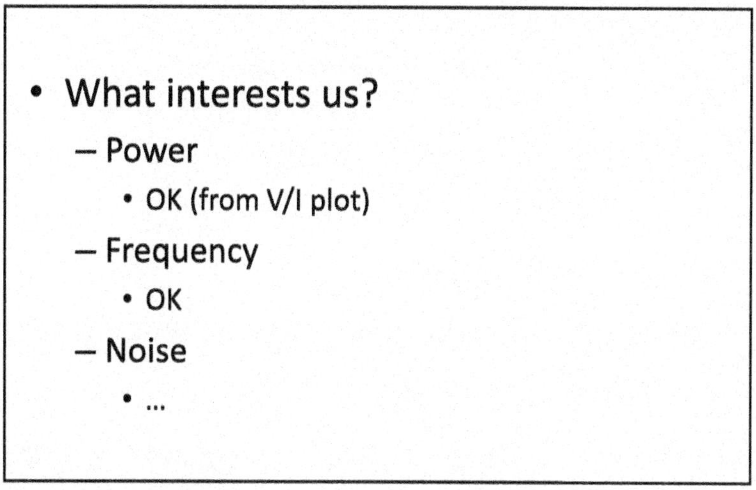

Fig. 4.14 VCO ring characteristics

Now we have covered two of the most important VCO characteristics. The first, the VCO power can be derived from the voltage vs. current plot (*I/V* characteristic). The second, the VCO frequency can be studied through computer simulations (either transient or as we recommend through PSS), as described on the previous figure. For use as a VCO-ADC the third important characteristic is the noise. In this book we advocate to study this as an input-referred noise and not as output phase noise. We will explain that noise can be traded for power as in any analog circuit (Fig. 4.14).

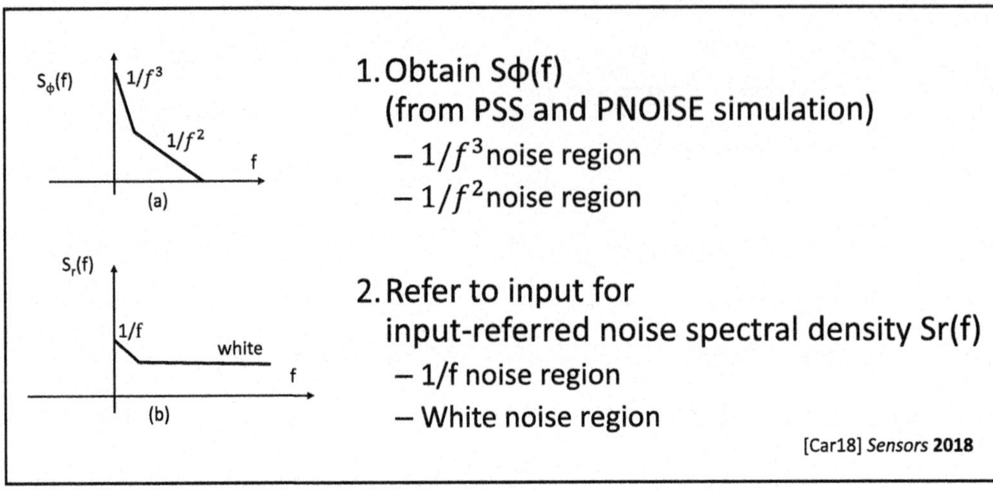

Fig. 4.15 Input-referred phase noise (see before)

Let us quickly review the recommended way to study noise in a VCO-ADC (see before, Figs. 2.7–2.9). First the output phase noise should be obtained from a combination of a PSS (Periodic Steady-State) and a PNOISE (periodic noise) simulation. This output phase noise will have the well-known asymptotic regions shown in Fig. 4.15a, following the well-known Leeson model [Abi06]. The most important regions are the $1/f^3$ and the $1/f^2$ regions (which correspond to input-referred flicker noise and white noise, respectively). Then in a second step we obtain the VCO sensitivity (K_V) from a PSS simulation. Finally, we can refer the output phase noise to the input by differentiating it and dividing it by the VCO sensitivity. This results in a plot such as the one shown in Fig. 4.15b, where we have a $1/f$ noise region and a white noise region.

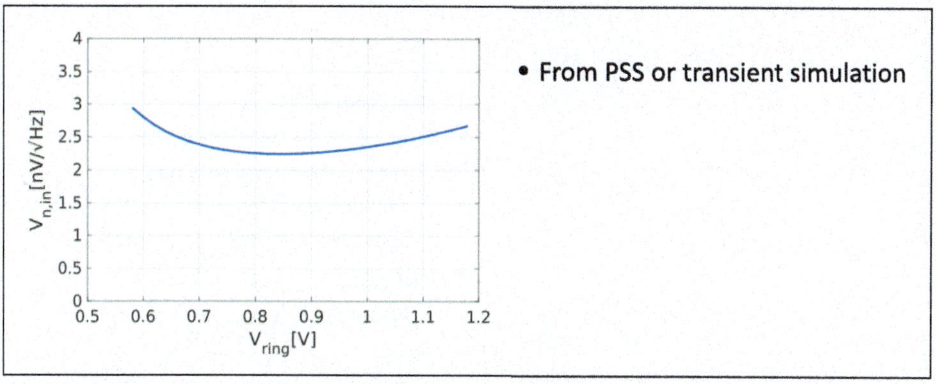

Fig. 4.16 Ring characteristic: input-referred white noise

The result of following the procedure described above for our example ring oscillator in 65 nm CMOS is shown above. Here, for every value of the ring voltage V_{ring} the input-referred phase noise spectrum has been determined and the white noise has been captured from the flat region in the noise spectrum plot. This experiment was then repeated for different values of the ring voltage to obtain the entire curve (Fig. 4.16).

From the plot it is clear that the input-referred white noise is quite flat over the entire represented V_{ring} range. In this case it is around 2.5 nV/$\sqrt{\text{Hz}}$ over most of the input range.

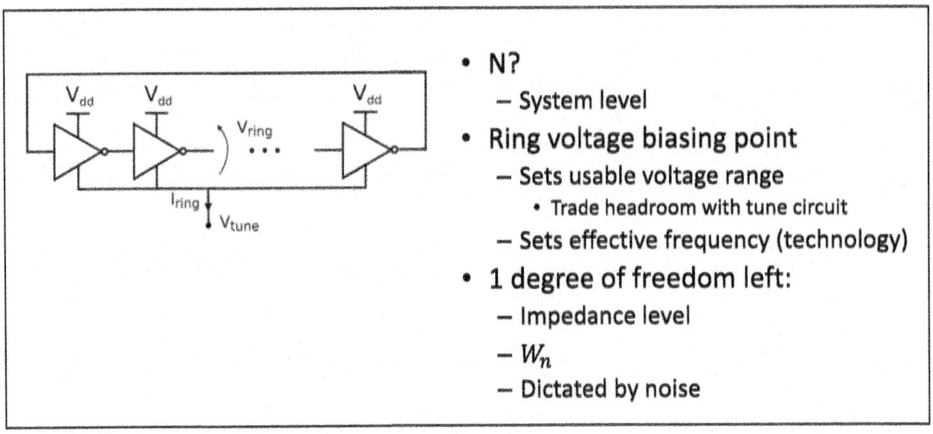

Fig. 4.17 VCO circuit: what is there to design?

Let us recapitulate. What is there to design in the VCO ring circuit?

First, it should be emphasized that N, the number of cells should be decided at the system level (taking the digital readout, etc. into account) (Fig. 4.17).

The next step is to decide the DC biasing point of the ring voltage V_{ring}. For this there are different considerations, but it is definitely clear that there is a trade-off with the headroom for the tune circuit. Especially at today's low supply voltages, this is an important concern. It also sets the effective frequency, see Fig. 4.13, and hence also the quantization noise for a given bandwidth and sampling frequency.

Keeping in mind that the inverters are simple CMOS inverters with a fixed ratio of the NMOS vs. the PMOS strength, after these considerations, there is only one degree of freedom left: i.e., the core inverter sizing. We will now focus on its W/L (basically its drive strength), which corresponds to the VCO's impedance level at the tune node.

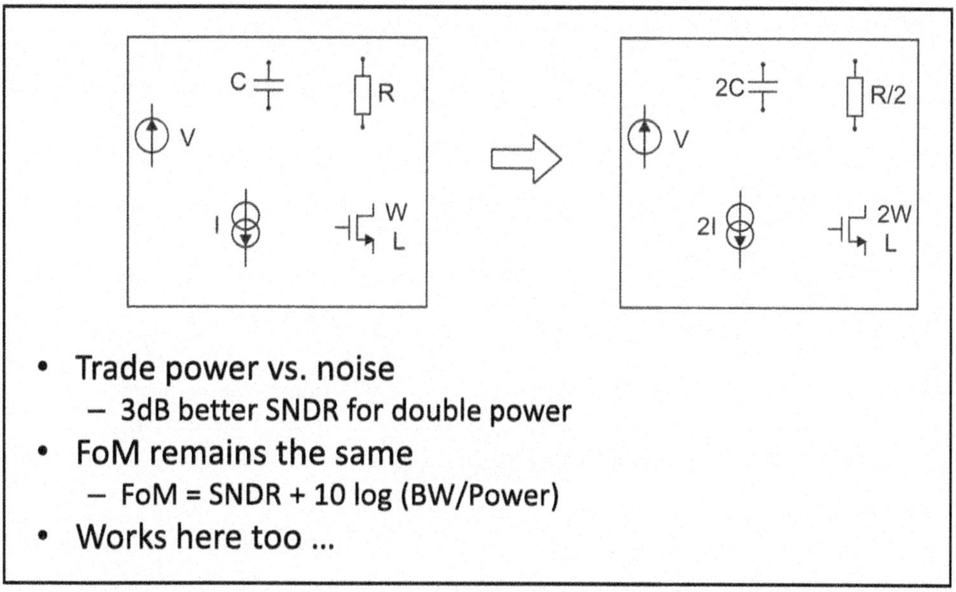

Fig. 4.18 Impedance scaling

In order to fully understand the meaning of the impedance level, we have to review a technique probably known to most designers: *impedance scaling*. This technique is represented in Fig. 4.18, where we start from a given design represented by the rectangle on the left. Let us assume that this is a (nearly) optimal design, which achieves all desired functionality with good power efficiency. In this design the signals are represented by voltages in the circuit. The circuit is assumed to be powered by a fixed supply voltage. Assume that a new version of the circuit is desired, this time with lower noise. For this, impedance scaling can be used. Since the original design was already optimal, we want to retain exactly the same voltage-mode behavior, since in the proposed circuit signals are represented by voltages. To achieve this we can make a new design where all impedances are scaled down with the same factor (in the figure a factor 2). If the same voltages are applied to the circuit, the internal voltages will remain the same. The current however will have increased with the same factor as the impedance scaling factor.

This well-known technique allows to trade power consumption vs. noise according to a factor of 2 power increase for an improvement of 3 dB in noise. A correct ADC Figure of Merit (FoM) should be invariant under impedance scaling [Rab97]. Some early ADC FoMs do not respect this invariance and hence were basically nonsensical. A well-known example of such an incorrect FoM is Walden's FoM [Wal99]. The most popular correct FoM is commonly called Schreier's FoM: FoM = SNDR + 10 log (BW/Power) [Sch05].

The key message is that this technique works on every analog circuit and hence also works here for ring oscillators. Hence, every ring oscillator design in principle can be scaled by the recipe shown in Fig. 4.18. A side note on this is that *everything* should be scaled: also all the capacitors. In a ring oscillator, a significant part of the capacitance can be due to parasitic interconnect capacitors, which however are difficult to scale accurately. Hence, in a practical design, after layout, the impedance scaled circuit is likely to have slightly different operating speed as the original version (either lower or higher). However, the 3 dB per factor 2 of power law is not affected by this effect.

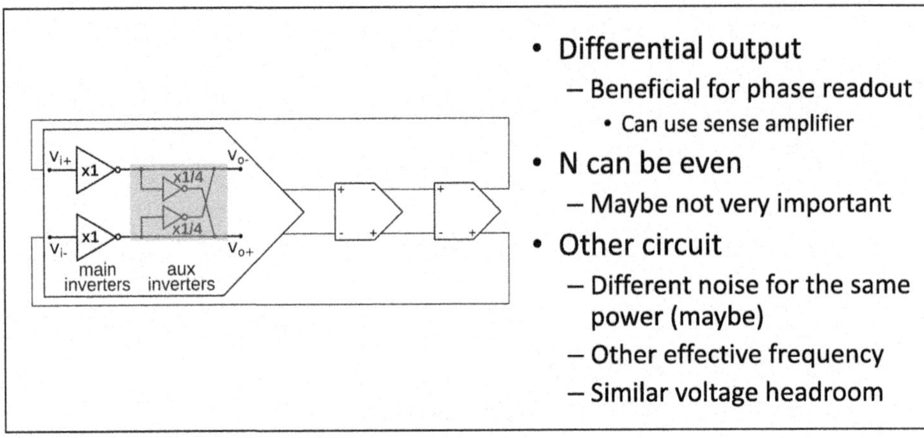

Fig. 4.19 Ring oscillators with differential delay cells

Until now we have been focusing on ring oscillators with straightforward inverters. But there are also a few important variants. A particularly important class are the differential delay cells. Even if we restrict ourselves to pseudo-digital circuits (built with *digital* inverters), there are two important variants. The first variant is shown on Fig. 4.19, where we can see a differential delay cell consisting of two strong *main inverters* and two weaker *auxiliary inverters* [Bab16], [Raz19]. There is no explicit common-mode feedback, but the auxiliary inverters create a balancing mechanism, which can be understood, for instance, by looking at the circuit as a Nauta cell [Nau92]. The sizing of the *auxiliary inverters* is a trade-off. If they are not sufficiently strong, the balancing mechanism might fail and in the worst case the oscillator may fail to start up. But if they are too strong, they increase the noise and the power consumption and lower the oscillation frequency. An established compromise is to size them a quarter of the strength of the *main inverters* [Bor19a], [Raz19].

A first advantage of this differential set-up is that the oscillation signal can now be read out as a differential signal. In this way, a sense amplifier (e.g., strong arm [Xu19]) can be used, allowing a more accurate phase detection and also less susceptibility to read out metastability (see below).

A second advantage of using differential delay cells is that the number of cells can now be even, because the correct sign in the feedback loop can always be obtained by cross-coupling the differential signal wires. In the context of VCO-ADCs this is not a major advantage, but it gives a little bit of extra design freedom.

Clearly this is a different circuit from the previous one, so the noise vs. power performance as well as the effective frequency is different compared to the previous circuit. The voltage headroom is similar.

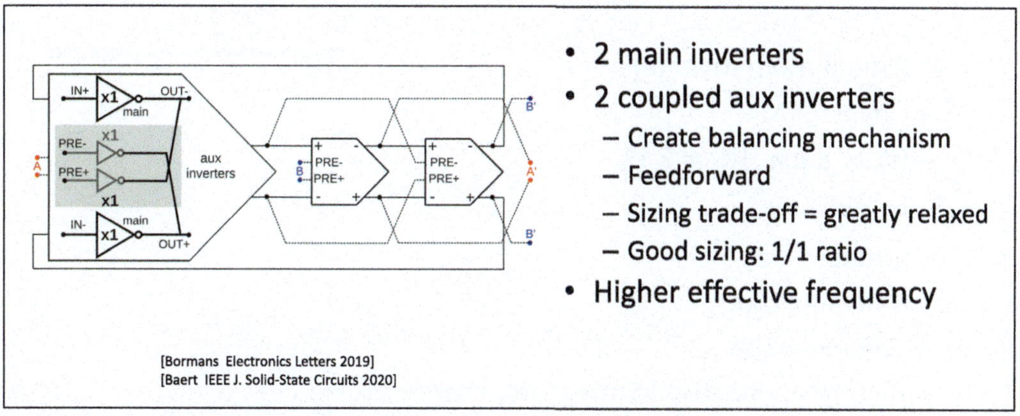

Fig. 4.20 Differential delay cell with feedforward cross-coupling (improved circuit)

The previous delay cell used auxiliary inverters with direct (in the cell) cross-coupling. A second important differential ring oscillator is based on a delay cell where the auxiliary inverters use feedforward cross-coupling [Kov18], [Bor19a], [Bae20]. The circuit for the case of a ring oscillator with three cells is shown in Fig. 4.20. Again, these auxiliary inverters create a balancing mechanism between the cell's differential outputs. In the study described in [Bor19a], it was found that the best sizing for this circuit is quite different from the previous circuit: it is best to size the auxiliary inverters simply equal to the main inverters, nearly completely eliminating the risk that this circuit would not start up.

In addition to the improved robustness, the main advantage of this circuit is that it has a significantly higher oscillation frequency compared to the delay cell with direct cross-coupling. This enables higher VCO-bandwidth for the same quantization noise (and hence the same SQNR) or better SQNR for the same bandwidth.

[Bormans IEEE Trans. Circuits & Systems I, 2021]

- Identical main inverter:
  - L minimum
  - $W_n = 8\ \mu m\ , W_p = 2\ W_n$
- Caution
  - Different area
  - Different Power
  - Different noise
  - Later fair comparison through impedance scaling.

Fig. 4.21 Comparison of three different unit cells (1) 65 nm CMOS case study

To further quantify the differences between the three circuits we will perform a comparative case study for 65 nm CMOS technology. In our case study we will use the same core (main) inverter in the three different unit cells (see Fig. 4.21).

This makes it easy to compare the circuits, but it also means that the three circuits will have a very different area, power, and noise (due to the different number of inverters).

Hence, the initial comparison is not a fair one in terms of power efficiency. To fix this we will make a fair comparison later on where all circuits are impedance scaled towards the same noise performance.

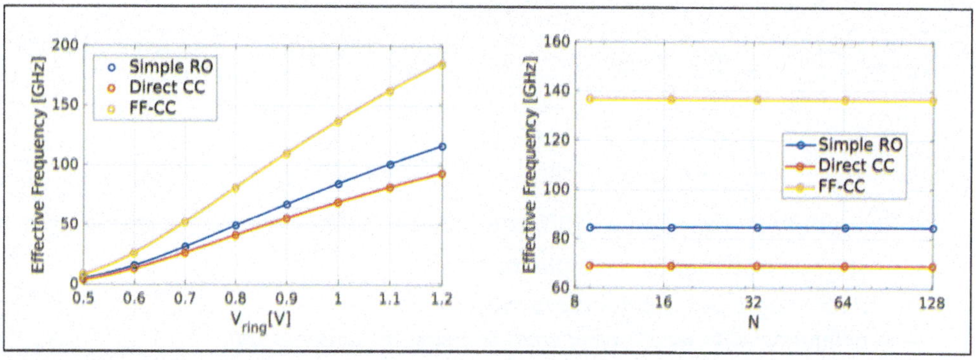

Fig. 4.22 Comparison of the three different unit cells (2) 65 nm CMOS case study

First the effective oscillation frequency has been studied. This study was done at the schematic level (remember the disclaimer on Fig. 4.3) and hence layout effects are not properly accounted for. But this is the same for the three cells and hence this effect should not affect the trends in the overall comparison. On the left plot in Fig. 4.22, we see the effective frequency as a function of the ring voltage V_{ring}. The simulation is carried out for the case of $N = 9$ cells in the ring. We can see that the improved differential circuit with feedforward cross-coupling consistently performs best. The traditional differential circuit with direct cross-coupling is worst and the simple (single-ended inverter) circuit is in between.

The theory predicts that the effective frequency is independent of the number of cells and this is confirmed by many more simulations. In particular, in the right on Fig. 4.22 we see a plot of the effective frequency for the case of a fixed ring voltage $V_{ring} = 1$ V, as a function of N, the number of cells in the ring, confirming this for all unit cells circuits.

Case 1: Reference case ($V_{ring} = 1V$)				
VCO Type	$W_{n,main}$ [μm]	f_{eff} [Hz]	P_{DC} [W]	$V_{n,in}$ [V/$\sqrt{Hz}$]
Simple	8	84.98G	2.08m	2.36n
Direct CC	8	69.67G	4.38m	1.88n
FF-CC	8	138.00G	12.53m	1.03n

Case 2: Scaled for equal input referred noise ($V_{ring} = 1V$)				
VCO Type	$W_{n,main}$ [μm]	f_{eff} [Hz]	P_{DC} [W]	$V_{n,in}$ [V/$\sqrt{Hz}$]
Simple	44.30	84.98G	11.52m	1n
Direct CC	28.26	69.67G	15.47m	1n
FF-CC	8.55	138.00G	13.39m	1n

- Best effective frequency = differential with feedforward cross coupling
- Best power efficiency = simple inverters
  - disclaimer comes later (phase readout circuit ...)
  - differential with feedforward cross-coupling close runner-up
- Differential with direct cross-coupling (traditional) = worst

Fig. 4.23 Comparison of the three different unit cells (3) 65 nm CMOS case study

Next, the input-referred has been studied through the approach described above. The resulting white noise for our case study is shown in the table on the left in Fig. 4.23. Here only the input-referred white noise component was considered. It can be seen that the noise is very different for the different cells. However, as expected, also the power consumption is very different, impeding an immediate comparison.

To make a comparison possible, the simulation results are re-scaled by applying impedance scaling towards an arbitrarily chosen noise level of 1 nV/$\sqrt{\text{Hz}}$. The corresponding results are shown in the table on the right in Fig. 4.23. Now we can immediately see that the simple circuit achieves this noise level with the lowest power and hence has the best power efficiency. The improved differential circuit (with feedforward cross-coupling) is the second best while the traditional differential circuit (with direct cross-coupling) has the worst power efficiency. It should be noted however that the differences in efficiency are relatively modest.

The tables in Fig. 4.23 also shows a column with the corresponding sizing of the main inverter, which ends up to be very different for the three circuits. But this should not be misunderstood, since the number of inverters in the three circuits is very different. After impedance scaling the overall area of the three circuits is similar.

The effective frequency was already studied before and does not change with the impedance scaling. Here, as also shown in the tables, the improved differential circuit (with feedforward cross-coupling) is clearly the best.

Based on these results there seem to be two sensible design options. If a high effective frequency is important, for example, when the bandwidth is pushed to the limit, then the differential circuit with feedforward cross-coupling should be used. If the focus is on power efficiency, then the simple circuit is the best. However, we have already mentioned that the differential circuits allow more accurate phase readout circuits (with a sense amplifier). Based on this, this conclusion could be partially amended.

- VCO analog performance mostly independent of number of unit cells, N
  - Except 1/f noise
- Quasi-static I_{ring} vs V_{ring} characteristic
  - Insight in voltage headroom
- Effective frequency vs V_{ring} characteristic
- Input-referred noise
  - Very low for given power
  - Good sensitivity!
- 3 possible circuits
  - Simple inverter: best power efficiency
  - Differential with feedforward coupling: best frequency
  - Differential with direct coupling (traditional): worst

Fig. 4.24 Wrap up

A summary of the circuit considerations for the ring circuit in VCO-ADCs can be found in Fig. 4.24.

Chapter 5
Analog VCO Coupling Stages: gm, Source Follower

G. Gielen et al., *Time-encoding VCO-ADCs for Integrated Systems-on-Chip*,
https://doi.org/10.1007/978-3-030-88067-5_5

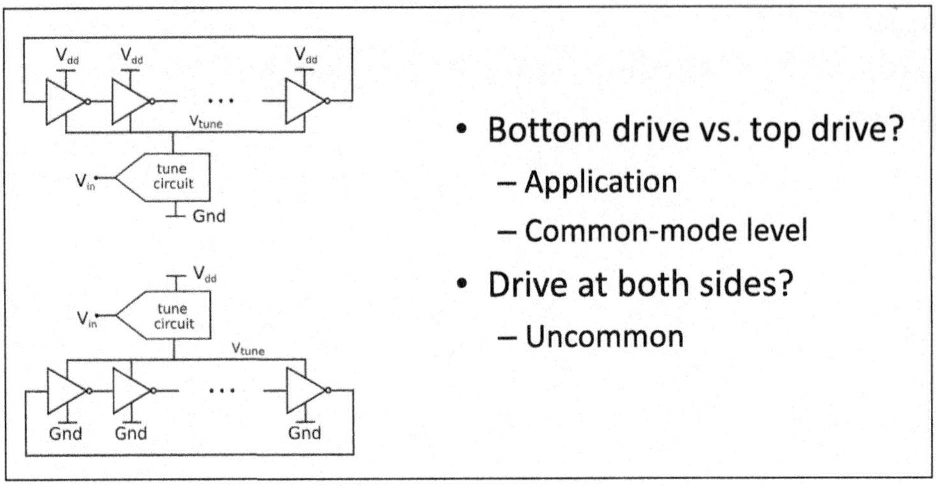

Fig. 5.1 Driving a ring oscillator (tune circuit)

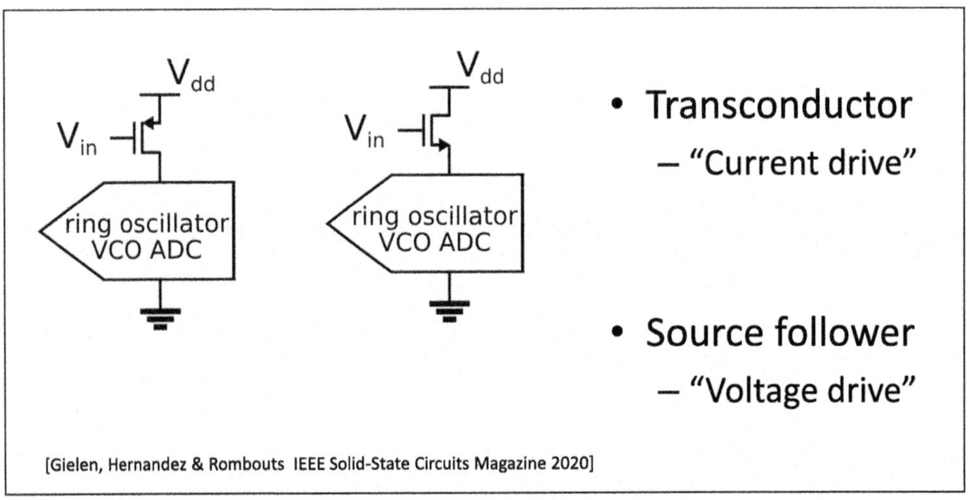

Fig. 5.2 Driving a ring oscillator (1)

Above we have been focusing on the ring itself, but equally important is the circuit to drive the ring. This is nearly always done through the "*supply lines*" of the inverters in the ring. Here, there are two options: either tune the "V_{SS}" (Gnd) line or alternatively tune the "V_{dd}" line. The former is called *bottom drive*, the latter *top drive* (Fig. 5.1).

Both options are viable and hence there is no obvious best choice. Very often there are application-level considerations that already rule out one of the two options. In particular the common-mode level of the input voltage (assuming a pseudo-differential setup) is different for both options. Hence, if the common-mode level is fixed by the application, this will already set the drive side.

In principle, it is also possible to drive the VCO at both sides. But in the context of VCO-ADCs, this seems to be uncommon.

Another design decision is on whether to drive the ring oscillator in "current mode" or in "voltage mode." The simplest thinkable current-drive circuit is a single-transistor transconductor, but much more sophisticated transconductors have been used as well (Fig. 5.2).

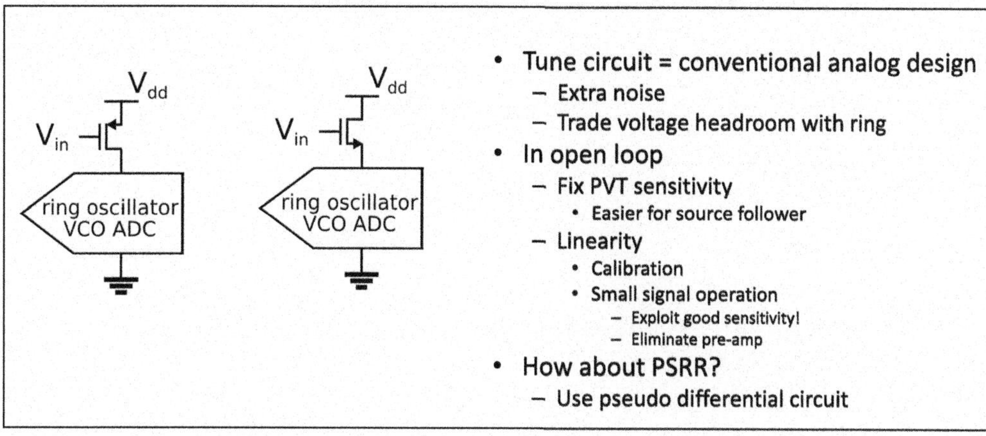

Fig. 5.3 Driving a ring oscillator (2)

The simplest voltage drive circuit is a single-transistor source follower.

The design of the tune circuit is quite conventional analog design. The main challenges are the extra noise that the tune circuit injects, which directly adds to the input-referred noise. Also, the tune circuit is in series with the ring. Hence, in a low-voltage design, the voltage headroom must carefully be partitioned between the tune circuit and the ring (Fig. 5.3).

If the VCO is used in a closed-loop configuration, these are the main constraints.

But if the VCO is used in an open-loop configuration, some more considerations should be taken into account.

First, measures should be taken to make the overall solution robust against PVT variations, which have a severe impact on the overall ADC gain. Solving this is definitely easier for the source follower than for the transconductor because the gain of a source follower (regardless of PVT variations) is basically 1, not leaving much room for variations. In contrast, the transconductance of a simple transistor can vary significantly.

Second, also the nonlinearity of the tune circuit directly contributes to the overall ADC nonlinearity and should be addressed. There are a number of solutions for this, such as calibration and the coarse fine structure (see Chap. 6 of this book). A usually overlooked but very viable solution is to exploit the good sensitivity of VCO-ADCs and simply operate the VCO-ADC in a small-signal regime where the linearity should be OK (see Fig. 3.3). At the system level this gives the advantage that the VCO pre-amplifier and driver can be eliminated, potentially greatly simplifying the overall system.

An important remark is that in their basic form all discussed ring oscillator circuits and the associate tune circuits suffer from poor power-supply rejection. But (as explained above, Fig. 3.2) this can be addressed by deploying the ring oscillator VCO-ADC in a pseudo-differential configuration. This pseudo-differential setup should always be the default, from which you only deviate if there are very good reasons to do so.

A noteworthy and unconventional tuning circuit is the resistive tuning circuit shown in Fig. 5.4. The bottom-drive configuration is represented, but also a top drive version is possible. It couples the input voltage through a resistive voltage divider. This circuit does not have the best noise performance due to the inherent attenuation over the two resistors. However its linearity performance is excellent with a reported linearity better than 12-bit [Bab16], [Bor19a], [Bor19b]. Compared to many other (open-loop) tuning circuits, this circuit has the additional disadvantage that the input impedance is relatively low. Hence an amplifier with sufficient drive strength is needed to drive the input resistive network.

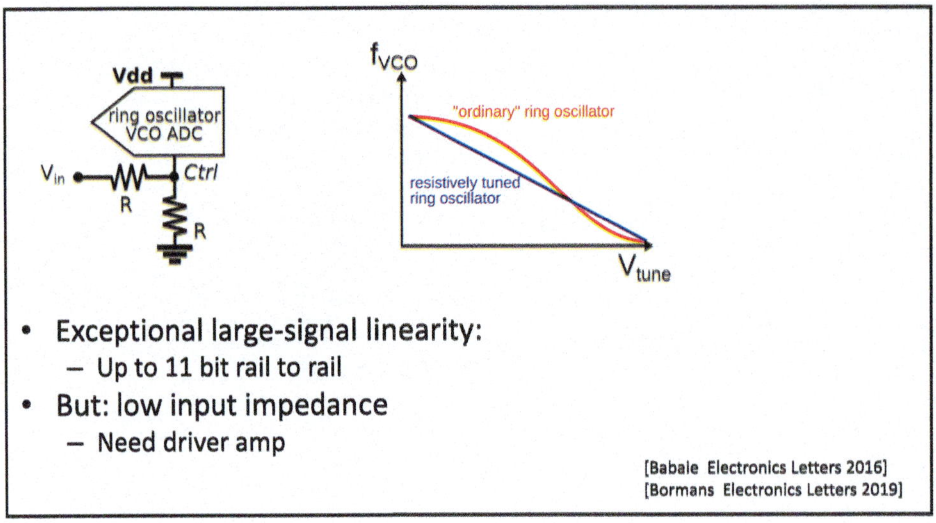

Fig. 5.4 High-linearity resistive tuning circuit

Until now, one aspect of the VCO has only vaguely been addressed, i.e., the (digital) buffers to read out every VCO phase. These buffers have to convert the VCO signals into full-scale digital waveforms.

However, the VCO signal is modulated by the input signal. This is illustrated in Fig. 5.5 where the VCO is driven at the top in voltage mode (through a source follower). The phenomenon is similar for bottom drive, as well as for top drive.

As can be seen, the voltage V_{ring} varies together with the input signal. Since the VCO output voltage oscillates between zero and V_{ring}, the VCO output voltage exhibits an amplitude modulation effect from the input signal.

The ideal crossover point for the readout buffer is mid-scale but this mid-scale voltage is modulated as well. This makes it difficult to design a buffer that has the right crossover point (there is not just one "right" crossover point). As a result of this, during a transition, the output of the buffer can stay relatively long in a poorly defined digital state (not really *high* or *low*). This in turn increases the probability of metastability in the actual sampling flip flop which is driven by the readout buffer. To avoid this, a designer can make a custom design of the readout buffer, to get a reasonable crossover point. This can be done by adjusting the relative strength of the NMOS and PMOS transistors in the digital buffer but also by combining low-threshold-voltage and high-threshold-voltage transistors (if they are available). But even then, typically multiple stages have to be cascaded and special low-aperture sampling flip-flops have to be used.

These problems can easily be overcome in a differential VCO, where the readout can be done by sense amplifier (e.g., a Strong Arm dynamic comparator). In this way the sampling and buffering can efficiently be combined with a very low probability of metastability. In this way the readout for a differential VCO might be more efficient than for a single-ended VCO.

In the case of a closed-loop VCO, there are significantly more options to drive the loop. An example is shown in Fig. 5.6 where the input signal is coupled into the ring through a coupling resistor and the feedback DAC is in current mode. This is definitely not the optimal choice in terms of VCO circuit noise: on top of the PMOS biasing current source, there are extra noise contributions from the input resistor as well as from the feedback DAC. However, it has the desirable property that the ring voltage V_{ring} is kept nearly constant due to the nullator operation of the loop.

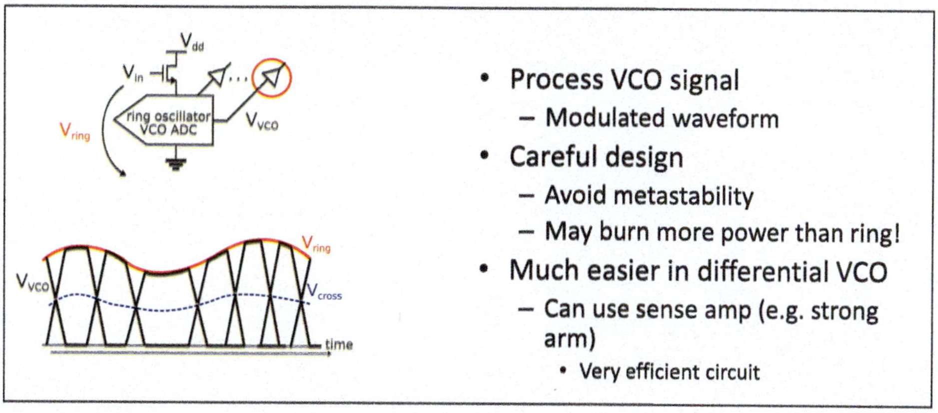

Fig. 5.5 Readout buffer pitfall

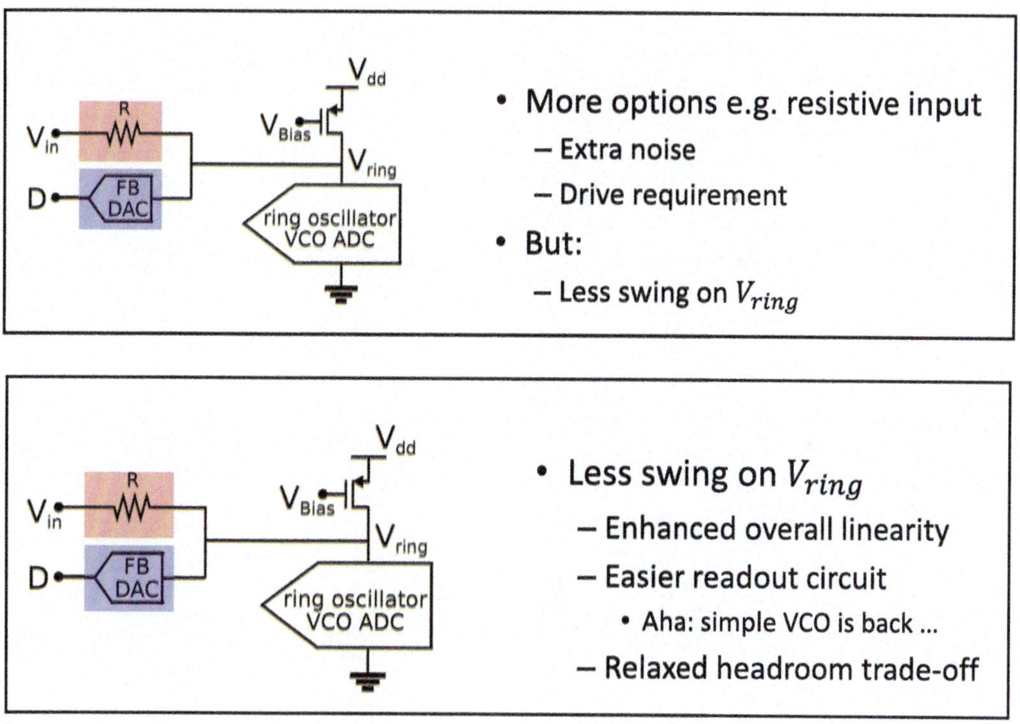

Fig. 5.6 Driving a ring oscillator in closed loop

With the reduced swing on V_{ring}, the effect of the VCO nonlinearity is greatly reduced (similarly as in every closed-loop feedback system). However, it also greatly reduces the modulation effect of the VCO voltage described in Fig. 5.5. Hence the associate problems with the readout circuit (described in Fig. 5.5) are largely eliminated. Therefore, in this case the simple inverter-based VCO may be the best design choice again. An additional benefit is that the trade-off between the drive circuit's and the VCO's headroom is greatly relaxed. In the next part such closed-loop architectures will be further elaborated.

Chapter 6
Closed-Loop VCO-ADC Architectures

G. Gielen et al., *Time-encoding VCO-ADCs for Integrated Systems-on-Chip*,
https://doi.org/10.1007/978-3-030-88067-5_6

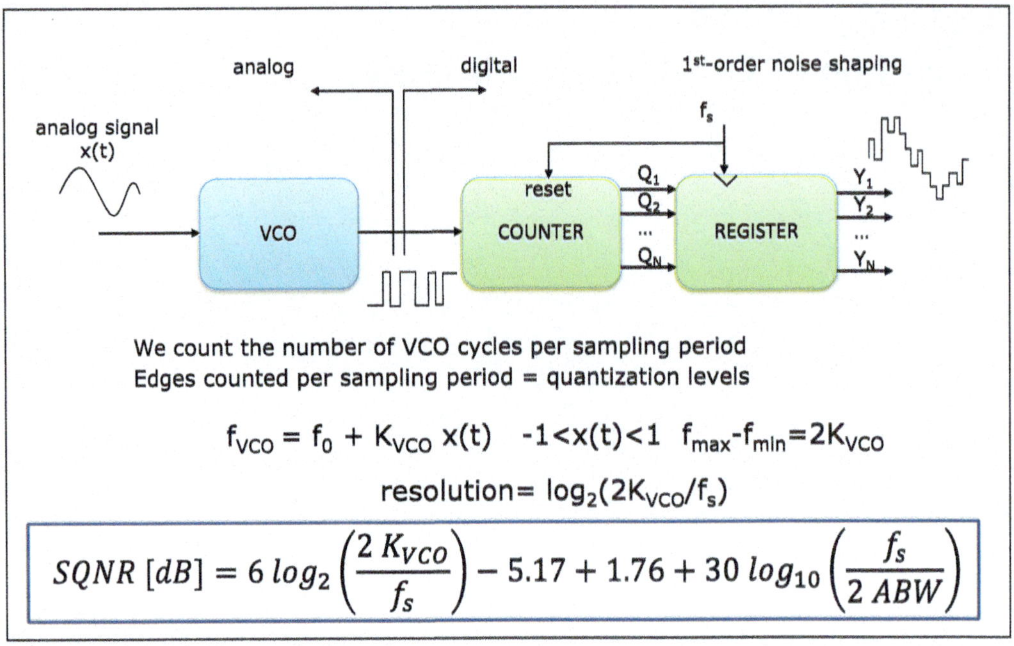

Fig. 6.1 SQNR of a VCO-ADC?

What is the peak signal-to-quantization-noise (SQNR) ratio of an open-loop VCO-ADC as shown on Fig. 6.1? As seen before, the architecture counts the number of VCO cycles (at f_{VCO}) per sampling period (at f_s) and stores it in the register. The edges counted per sampling period correspond to the quantization levels. So the full range corresponds to the full frequency excursion $f_{max} - f_{min}$ of the VCO, which is equal to K_{VCO} times the full input swing of the VCO input signal $x(t)$. ($x(t)$ is normalized between −1 and +1 in Fig. 6.1.) Therefore, the resolution of this VCO-ADC is given by $\log_2(2K_{VCO}/f_s)$.

Assuming a maximum input signal, and using the equation for first-order noise shaping (due to the implicit integration in the VCO), the expression for the peak SQNR [in dB] becomes (see also Fig. 1.12):

$$\text{SQNRpeak}\left[\text{dB}\right] = 6\log_2\left(\frac{2\,K_{\text{VCO}}}{f_s}\right) - 5.17 + 1.76 + 30\log_{10}\left(\frac{f_s}{2\,\text{ABW}}\right)$$

where ABW is the analog signal bandwidth. The ratio $f_s/(2^*\text{ABW})$ is the oversampling ratio OSR. The choice of f_s relative to ABW and f_{VCO} allows to increase the peak SQNR, as described in previous figures.

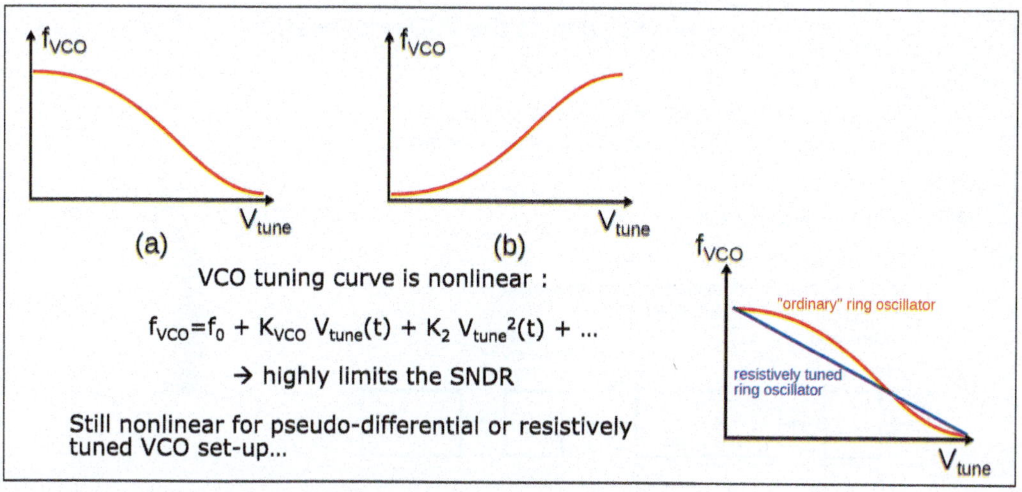

Fig. 6.2 Problem of VCO's nonlinearity

VCO-ADCs however have another big design challenge, and that is the nonlinearity of the VCO characteristic. A typical VCO tuning curve is either downward or upward nonlinear as a function of the input tuning signal V_{tune}, as shown in Fig. 6.2. Therefore:

$$f_{VCO} = f_0 + K_{VCO}\, V_{tune}(t) + K_2\, V_{tune}^{\;2}(t) + \ldots$$

where V_{tune} is the VCO input voltage, K_{VCO} is the linear conversion gain of the VCO, K_2 is the second-order nonlinearity coefficient, K_3 the third-order nonlinearity coefficient, etc. In differential implementations K_3 is often the dominant nonlinearity term.

It is important to emphasize that even when using pseudo-differential or resistively tuned (see previous figures) VCOs, the nonlinearity may improve but does not disappear completely.

This nonlinearity of the VCO tuning curve in open loop causes distortion, therefore limiting the SNDR and effective resolution (ENOB) of the VCO-ADC in practice.

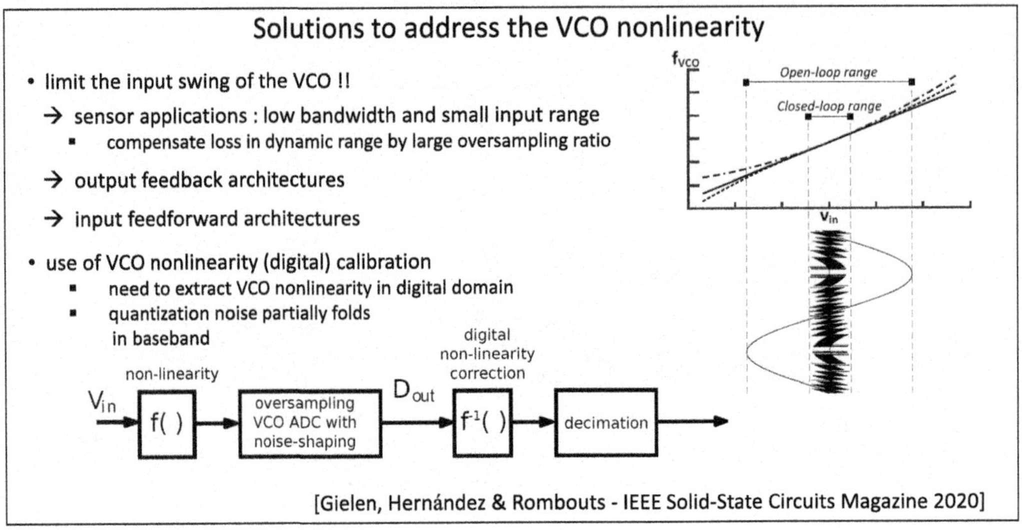

Fig. 6.3 Solutions to address the VCO nonlinearity

So how can we address the problem of the VCO nonlinearity when used for applications with a high SNDR or SFDR requirement?

There are basically two different ways to approach this problem [Gie20b]:

1. Limit the input swing of the VCO, so that the impact of the nonlinearity is smaller (see Fig. 6.3 top right). We will explain below how this can be done.
2. Use (digital) nonlinearity calibration to calibrate away the impact of the VCO nonlinearity on the ADC output (see Fig. 6.3 at the bottom).

So how can the input swing of the VCO be limited (see Fig. 6.3 at the top right)? This can be done in different ways.

(a) Applications like sensing typically have a low bandwidth and small input range, therefore limiting the nonlinearity problem, while a high dynamic range can still be obtained by using a large oversampling ratio f_s/ABW.
(b) The use of (negative) feedback in a closed-loop architecture also limits the input range below the open-loop range, hence limiting the impact of the nonlinearity and therefore reducing the distortion. These closed-loop architectures will be described in the subsequent figures.
(c) Also input feedforward architectures limit the input swing and therefore the impact of distortion, as will be discussed on the next figure.

The second approach is to use (digital) calibration of the VCO nonlinearity. The basic concept is shown in Fig. 6.3 at the bottom. If the VCO nonlinearity $f()$ can be extracted in the digital domain, then this information can be used to carry out the inverse nonlinearity correction $f^{-1}()$ in the digital domain, therefore largely cancelling the nonlinearity. Different practical implementations of this concept exist, and care must be taken as the quantization noise can partially fold into the baseband [Gie20b].

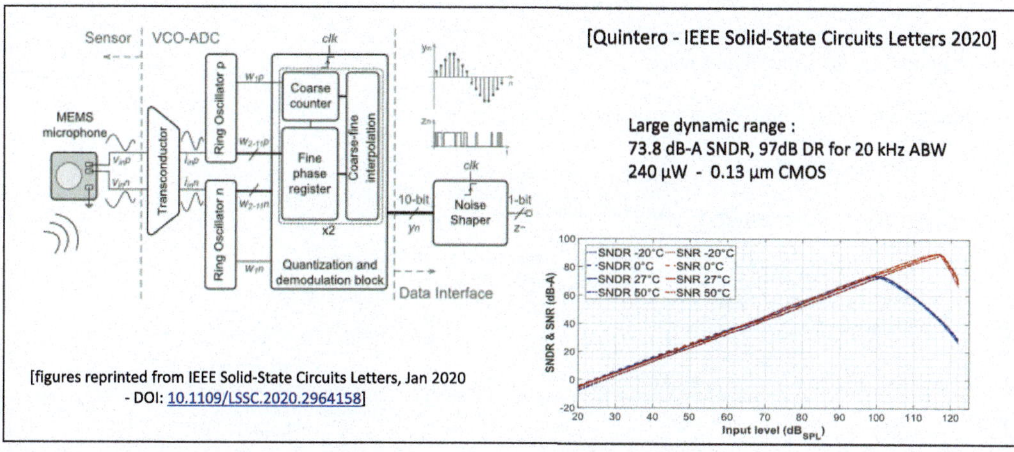

Fig. 6.4 Example: counting architecture with small input range

Figure 6.4 shows a recent example of a mostly digital VCO-ADC with a small input range by interfacing directly to a capacitive MEMS microphone [Qui20]. The ADC structure is shown in Fig. 6.4. It is based on two ring oscillators (ring oscillator p and ring oscillator n) and coarse-fine counting circuitry. The coarse counter and the fine counter (edge-triggered phase register) are synchronized using a data scrambling technique in the coarse-fine interpolation circuit to mitigate metastability and timing errors. This reduces the impact of asynchrony errors that otherwise would be audible in the output. The method enables a very low power consumption in the digital post-processing circuit.

The circuit has been prototyped in 130 nm CMOS. It achieves 73.8 dB-A of peak signal-to-noise-and-distortion ratio (SNDR) and 97 dB of dynamic range (DR) in a bandwidth of 20 kHz, while consuming 240 µW from the 1.5 V/1.2 V power supplies and occupying an area of 0.04 mm^2. The Schreier Figure of Merit (FoM$_S$) is 176.2 dB. In a reduced power mode (bandwidth of 8 kHz) with relaxed oscillation parameters, it reaches 66.4 dB-A of peak SNDR and 93 dB of DR with a power consumption of only 77 µW.

The plot on the right in Fig. 6.4 shows the measured SNR (in red) and SNDR (in blue) (both in dB-A) as a function of the input level in normal mode, for different temperatures between −20 °C and +50 °C. The chip has a very high dynamic range DR of 93 dB. The reduction in SNDR due to distortion for larger amplitudes is shown clearly. The variation across the temperature range is less than 3 dB-A. The measured power supply rejection ratio (PSRR) is above 80 dB for both the analog and digital power supplies.

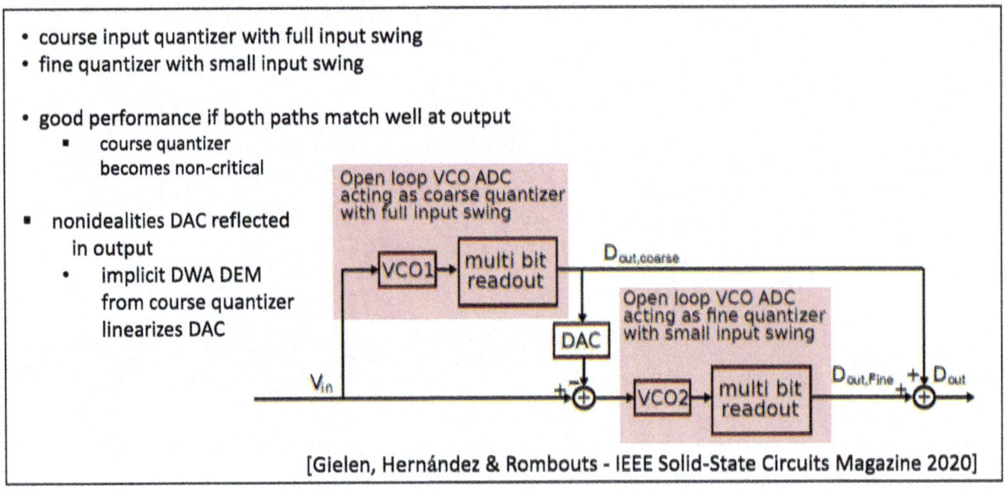

Fig. 6.5 Open-loop input feedforward architecture

Figure 6.5 shows the basic principle of the open-loop input feedforward architecture, which is one way to limit the input range and therefore the impact of the VCO nonlinearity, as described in Fig. 6.3.

As shown in the basic circuit block diagram on Fig. 6.5, the ADC is split in a first coarse quantization with VCO_1 and a second fine quantization with VCO_2. Since the output of the coarse VCO_1 is converted back to analog through the DAC and then subtracted from the input signal V_{in}, the input swing of the fine VCO_2 is small, hence causing little distortion. The outputs of the coarse VCO, $D_{out,coarse}$, and of the fine VCO, $D_{out,fine}$, are combined digitally at the overall output D_{out}.

Even though VCO_1 has the full signal swing at its input, if the path from VCO_1 through the DAC and VCO_2 to D_{out} is matched (digitally) with the direct path from VCO_1 to the output D_{out}, then the impact of the coarse quantizer becomes less critical. As a drawback of the structure, all nonidealities of the DAC are reflected in the output D_{out}. However, the output of the coarse quantizer $D_{out,coarse}$ is encoded with implicit dynamic element matching (DEM), hence also linearizing the DAC.

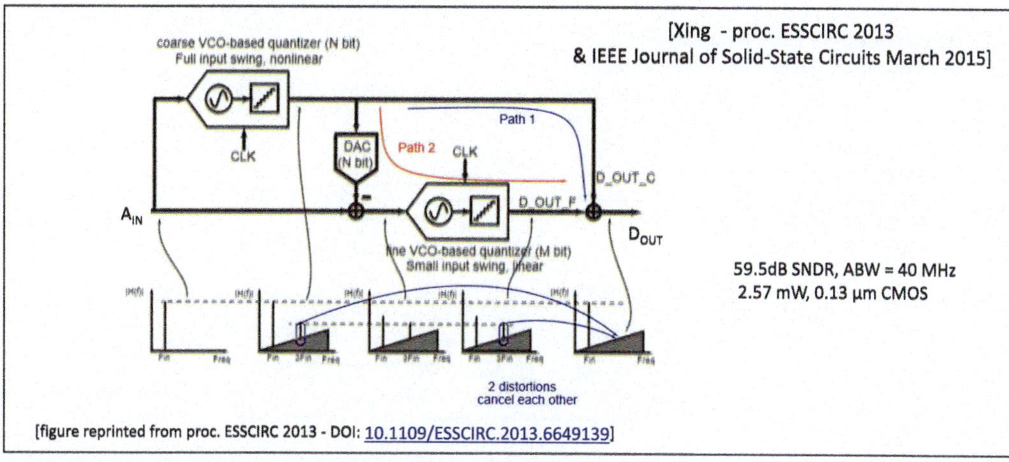

Fig. 6.6 Example: coarse-fine feedforward VCO-ADC topology

Figure 6.6 shows one example chip implementation [Xin13], [Xin15] of such a coarse-fine feedforward VCO-ADC structure in 40 nm CMOS. It is a two-step VCO-based ΔΣ ADC with first-order noise shaping and intrinsic nonlinearity mitigation. It has 40 MHz of analog bandwidth and 10 bits of resolution. With its open-loop structure and highly digital building blocks, a robust performance, high bandwidth, and high power efficiency are achieved. The nonlinearities of the coarse and the fine VCO-based quantizers are mitigated by distortion cancellation and voltage swing reduction schemes, respectively, as explained in Fig. 6.5. Because of the intrinsic DEM of the coarse VCO-based quantizer output, the matching requirement of the DAC cells is greatly relaxed.

Figure 6.6 also shows the signal spectra at the different nodes of the structure. The output of the coarse quantizer shows distortion, which however is also forwarded as input to the fine quantizer, and therefore largely cancelled when the two quantizer outputs are combined at the overall output D_{OUT}.

The design has been implemented in 40 nm CMOS. The experimental results showed that, with a sampling frequency of 1.6 GHz, the ADC reaches 59.5 dB of SNDR and 67.7 dB of SFDR for a bandwidth of 40 MHz. The power consumption is only 2.57 mW at 0.9 V power supply, corresponding (at its time of publication) to an excellent ADC Walden Figure of Merit (FoM_W) of 42 fJ per conversion step for a high-bandwidth ΔΣ ADC in 0.13 µm CMOS.

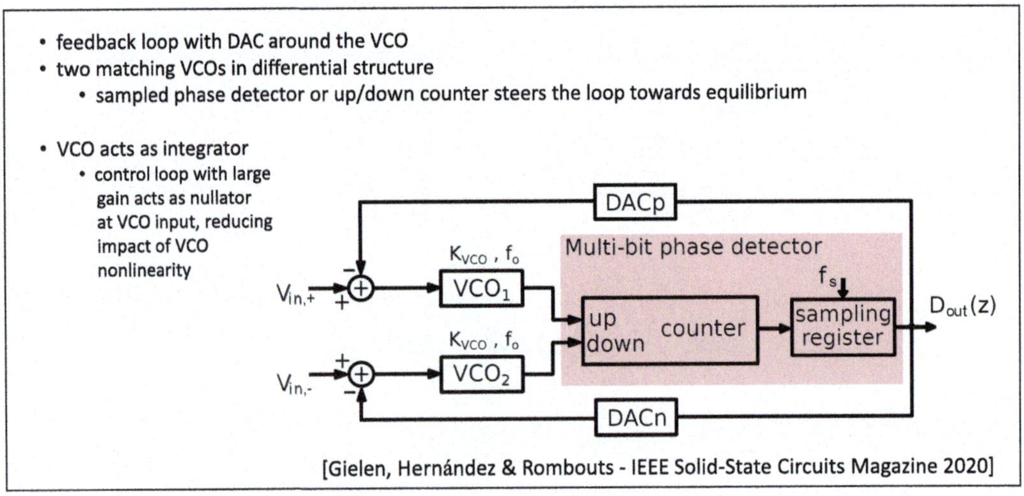

Fig. 6.7 Closed-loop output feedback architecture

The second way to reduce the impact of the nonlinearity of the VCO tuning curve in a VCO-based ADC is to feed back the digital output D_{out} through a digital-to-analog converter (DAC) and to subtract it from the analog input V_{in}, hence creating a closed-loop architecture. As discussed before, this negative feedback limits the input signal swing, hence suppressing the impact of the VCO nonlinearity and the resulting distortion.

As shown on Fig. 6.7, the architecture can be implemented in a differential structure with two matching VCOs. An up/down counter with sampling register (or alternatively a sampled phase detector) with the two VCO outputs as inputs steers the loop towards equilibrium.

Note that the VCOs act as integrator: the phase of a VCO output signal is the integral of its frequency. Therefore, this architecture acts like a negative feedback control loop with a large gain, hence creating a nullator at the input of the VCOs, which reduces the impact of their nonlinearity.

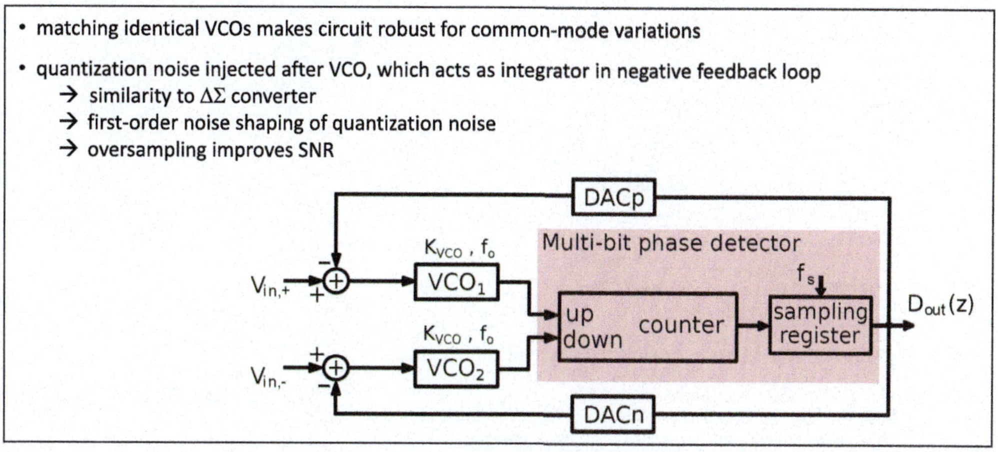

Fig. 6.8 Properties of closed-loop architecture

Besides the reduction of the distortion, this differential closed-loop structure also has some other interesting properties, which are described below (Fig. 6.8).

Firstly, considering the differential structure and assuming that the two VCOs (VCO_1 and VCO_2) ideally are identical, the circuit is very robust to any kind of common-mode variation that affects both VCOs in the same way (e.g., supply voltage variations, temperature variations, etc.), provided that the two VCOs match well. This may also explain the high PSRR values typically measured in this kind of VCO-ADC structures, like, for instance, in [Qui20]. Of course, as will be discussed later on, in reality mismatches between the VCOs exist, resulting in errors that may need to be addressed (depending on the requirements of the application).

Secondly, since the VCOs act as integrators and the up/down counter (or the phase detector) takes the phase difference between the two VCO outputs as input, this structure is functionally similar to a ΔΣ converter, be it in the phase domain rather than the voltage domain. This also implies that the structure intrinsically exhibits first-order noise shaping when oversampling the clock frequency f_s, hence improving the SQNR. This is particularly interesting for low-bandwidth applications like sensing, where the SNDR requirements can be very high, but where the input signal bandwidth is limited (at most typically some 20 kHz), hence easily allowing oversampling, while distortion is limited due to the feedback.

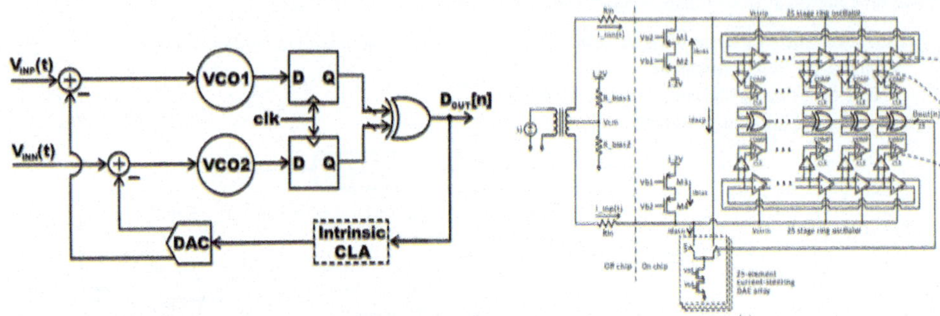

(a)

- 66.5 dB SNDR, ABW = 2 MHz, 1.8 mW, 0.03 mm^2 in 130nm CMOS

[Lee - IEEE JETCAS 2015]

[figures reprinted from IEEE Journal on Emerging and Selected Topics in Circuits and Systems, December 2015 - DOI: 10.1109/JETCAS.2015.2502166]

Fig. 6.9 Example design: first-order closed-loop VCO-ADC

Figure 6.9 shows an example of a first-order closed-loop ΔΣ VCO-ADC [Lee15]. As shown in the circuit diagram on the left, it uses the two VCOs as both quantizer and integrator, thus avoiding the use of power-hungry scaling-unfriendly OTAs and precision comparators. The two VCOs are arranged in a differential way, cancelling out even-order distortions. The XOR determines the phase difference between the two VCO outputs. The implementation with ring VCOs results in a natural rotation of the DAC selection elements, creating an inherent mismatch shaping capability (indicated as intrinsic clocked averaging or CLA) that automatically scrambles any mismatches in the feedback DAC. The actual circuit schematic is shown on the right in the Fig. 6.9.

A prototype chip of this ΔΣ ADC has been implemented in 130 nm CMOS, occupying a small area of only 0.03 mm^2 and achieving 66.5 dB of SNDR over a bandwidth of 2 MHz, while sampling at 300 MHz. The prototype consumes 1.8 mW from a 1.2 V power supply. The chip can also operate with a low analog supply of 0.7 V, achieving 65.8 dB of SNDR while consuming 1.1 mW of power. The highly digital structure offers good scalability perspectives towards advanced CMOS technologies.

- different VCO stages with feedback to increase the order of noise shaping
- highly digital implementations without typical analog blocks (opamps, comparators...)

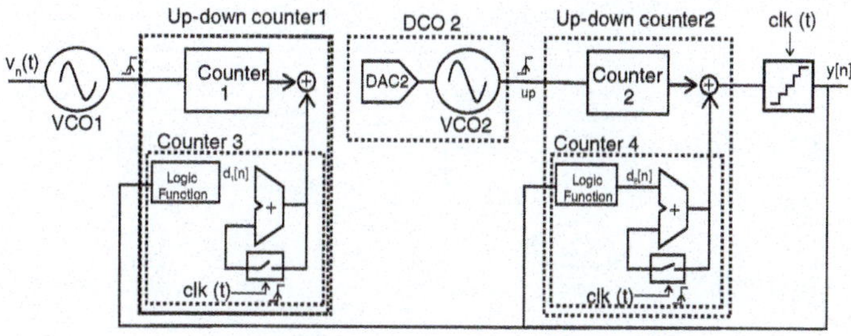

[figure reprinted from IEEE Journal of Solid-State Circuits, June 2018 - DOI: 10.1109/JSSC.2018.2799938]

[Cardes - IEEE Journal of Solid-State Circuits 2018]

Fig. 6.10 Example design: second-order closed-loop VCO-ADC

The next two figures show two examples of closed-loop VCO-ADCs with a higher order of noise shaping. Both use multiple VCO stages in cascade to achieve this, and both are implementations without traditional analog blocks (opamps, comparators).

The circuit on Fig. 6.10 [Car18a], [Car18b] is a second-order VCO-based ΣΔ ADC for audio applications. The highly digital converter is implemented using only oscillators and digital circuitry, without operational amplifiers nor other highly linear circuits, resulting in a compact area. The first VCO is outside the feedback loop, but the second VCO is inside the loop. The output is fed back to both the first and the second up-down counter. The circuit has been prototyped in a 0.13 μm CMOS technology. It occupies an active area of 0.02 mm^2, consumes 155 μA from a 1.8 V supply, and achieves an A-weighted dynamic range (DR) of 98 dB-A. A pseudo-differential configuration of two twin converters achieves 103 dB-A of A-weighted DR at the expense of doubling the area and power consumption.

- different VCO stages with feedback to increase the order of noise shaping
- highly digital implementations without typical analog blocks (opamps, comparators...)

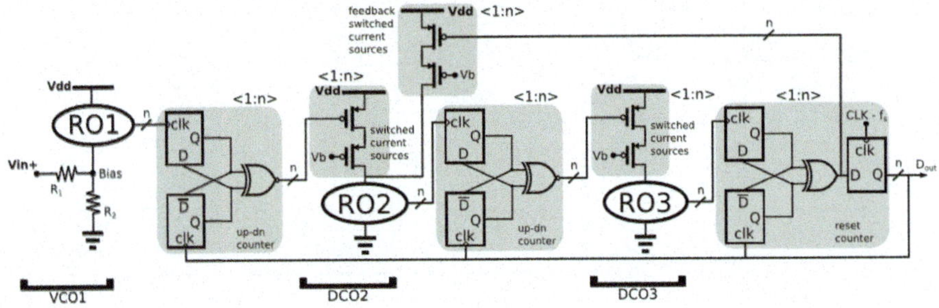

[figure reprinted from IEEE Journal of Solid-State Circuits, Aug 2017 - DOI: 10.1109/JSSC.2017.2688364]

[Babaie-Fishani – IEEE Journal of Solid-State Circuits 2017]

Fig. 6.11 Example design: third-order closed-loop VCO-ADC

As second example, the circuit on Fig. 6.11 [Bab17] is a mostly digital VCO-based continuous-time $\Sigma\Delta$ modulator with third-order quantization noise shaping. It is based on the combination of a VCO and a digital counter, which is shown to function as a continuous-time integrator. The counter consists only of digital blocks, and the VCOs are implemented using ring oscillators (RO1 to RO3). The design is digital-friendly and uses no traditional analog blocks like opamps, OTAs, or comparators. A prototype chip of a third-order VCO-based CT-$\Sigma\Delta$M for a bandwidth of 10 MHz has been developed in a 65 nm CMOS technology. It has a measured performance of 71 dB of dynamic range and 62.5 dB of SNDR for a bandwidth of 10 MHz. It consumes 1.8 mW from a 1.0 V analog supply and 1.9 mW from a 1.2 V digital supply. The first VCO is outside the feedback loop and uses the resistive signal drive techniques to improve the linearity (see Fig. 5.4 in this book). With digital calibration, the nonlinearity is pushed below the noise level, leading to an improved peak SNDR of 66 dB.

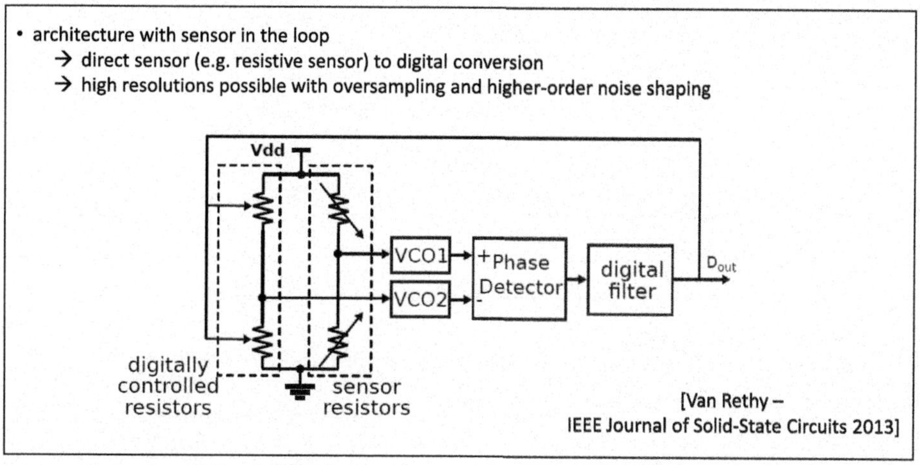

Fig. 6.12 Advantage for sensor interfaces

An extra advantage when using closed-loop VCO-based architectures for sensor readout is that the sensor can be included within the loop. This enables readout architectures with direct sensor to digital conversion. Figure 6.12 shows the architecture for the readout of a digital resistive sensor, that is put in Wheatstone-bridge configuration [Van13]. The differential resistive sensor is placed in one branch of the bridge, and digitally controlled resistors are placed in the other branch. The two midpoints of the two bridge branches are forwarded as controlling inputs to two (otherwise identical) voltage-controlled oscillators (VCO1 and VCO2). The oscillation waveforms of the two VCOs are compared in the phase detector. Its output is proportional to the phase difference between VCO1 and VCO2. The phase detector output is then sent to a digital filter that generates the final digital output. This output is also fed back to control the differential resistors in the left branch in the bridge. Due to the feedback configuration, equilibrium is reached when the two VCOs are in phase, i.e., when they run at the same frequency. The digitally controlled resistors then must have equal values to the actual physical resistors; therefore, the digital output must be the digitized version of the analog signal sensed by the resistive sensors.

As described in Fig. 6.8 this architecture acts as a $\Delta\Sigma$ modulator in the time (or phase) domain. Hence, oversampling and noise shaping (higher order is possible through the additional digital filter) increase the resolution. In addition, the architecture is resilient to common-mode variations that affect the two VCOs in the same way. The chip presented in [Van13] shows, for instance, excellent supply noise resilience.

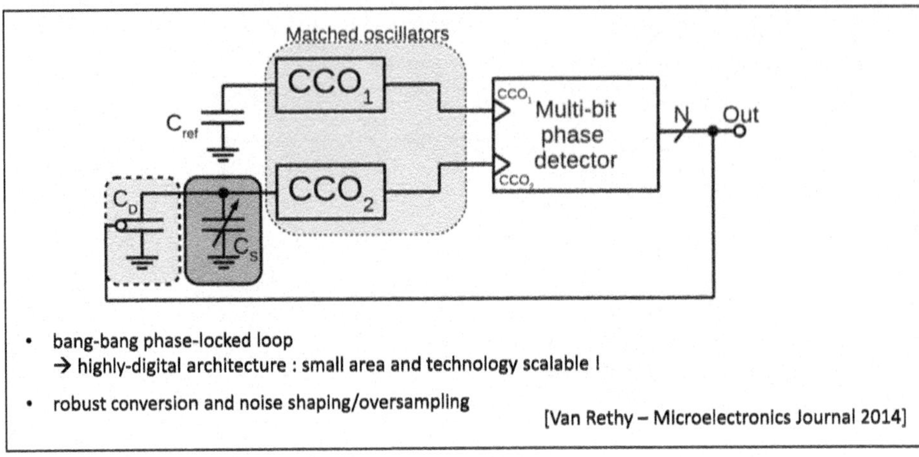

Fig. 6.13 Example design: capacitive sensor interface (1)

The same closed-loop VCO-based ADC concept can also be applied to the direct digital readout of capacitive sensors, i.e., sensors that measure the physical signal (e.g., pressure) as the variation of a capacitance. Figure 6.13 [Van14] shows a bang-bang phase-locked-loop-based readout architecture where the sensor capacitance C_S is driving the bottom "capacitance-controlled oscillator" (CCO). This could be a ring oscillator where the driving sensor capacitance is inserted as extra load for one of the stages of the CCO, hence changing the delay of that stage and therefore the oscillation frequency of the CCO. Note that the digitally switchable capacitance C_D is inserted in the same CCO_2 as the sensor capacitance, while CCO_1 on top has a fixed capacitive load C_{ref}. This ensures that CCO_1 runs at a fixed frequency, while the feedback loop makes CCO_2 run (on average) at this same fixed frequency. As both oscillators remain around a fixed operating point on the CCO capacitance-to-frequency characteristic, the impact of this characteristic's nonlinearity is highly reduced. To make this happen, the digitally switchable capacitance C_D will actually be driven such that it compensates any fluctuation of C_S: extra capacitance is switched on in C_D if C_S decreases (depending on the signal being sensed), and capacitance will be switched off in C_D if C_S increases.

The chip implementation in [Van14] uses a multi-bit phase detector for the phase comparison. The entire structure is highly digital, consumes a small area, and has high technology scalability. To the extent that the two oscillators are matched, the conversion is very robust and resilient to common-mode influences. The intrinsic first-order noise shaping in combination with oversampling results in an increased resolution. The prototype chip has been implemented in 130 nm CMOS, has an active area of only 0.0725 mm^2, and operates at a supply voltage V_{DD} as low as only 0.3 V, with a power consumption of only 270 nW. Later innovations to the basic architecture shown above have demonstrated how to improve the performance even further [Gie16].

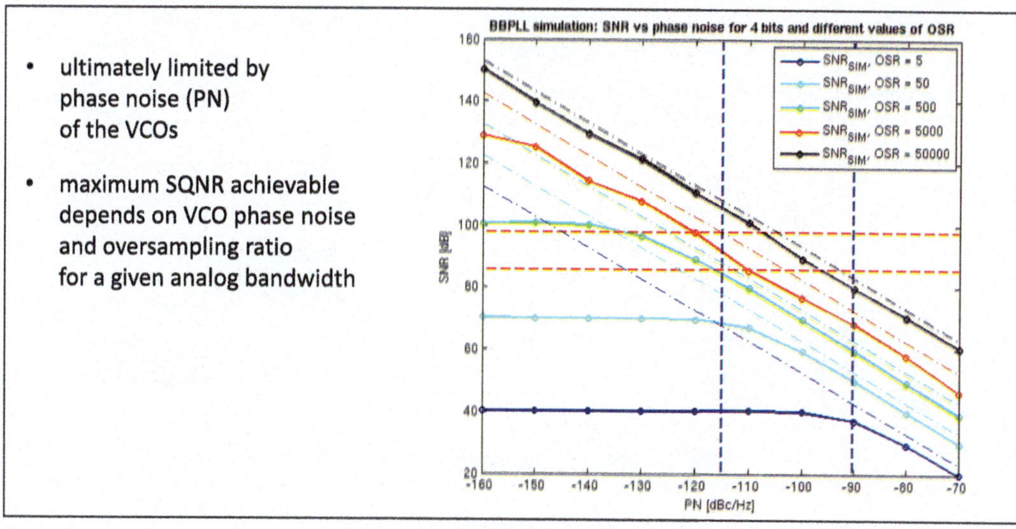

Fig. 6.14 Example design: capacitive sensor interface (2)

Since these architectures have intrinsic first-order noise shaping and possibly higher-order noise shaping, the use of oversampling improves the resolution. What is then the limit to their performance? Figure 6.14 shows that the SNR is ultimately limited by: (1) the quantization noise (which depends on the amount of oversampling and the order of noise shaping), and (2) the device noise, which manifests itself as phase noise (or jitter) in the VCOs (indicated on the x-axis as "PN" in dBc/Hz at some fixed offset frequency from the carrier). Figure 6.14 shows simulated maximum achievable SNR results for different values of oversampling ratio (OSR), ranging from OSR = 5 (blue curve) to OSR = 50,000 (black curve). At low amounts of phase noise (left part of the figure), the quantization noise dominates and limits the achievable maximum SNR to a flat level depending on the OSR value. At high amounts of phase noise (right part of the figure), the phase noise dominates and the achievable maximum SNR decreases with increasing phase noise.

So what performance can be achieved? Let us look at some typical numbers. If, for instance, an OSR of 500 (green curve) is used (which is acceptable for a low-bandwidth sensing application), a VCO phase noise PN between −110 and −120 dBc/Hz results in a maximum achievable SNR between 80 and 90 dB. Figure 6.14 of course is based on system-level simulations that need to be confirmed in actual circuit implementations and measurements, but that at least indicate the performance playing field that can be reached with these VCO-based ADC architectures.

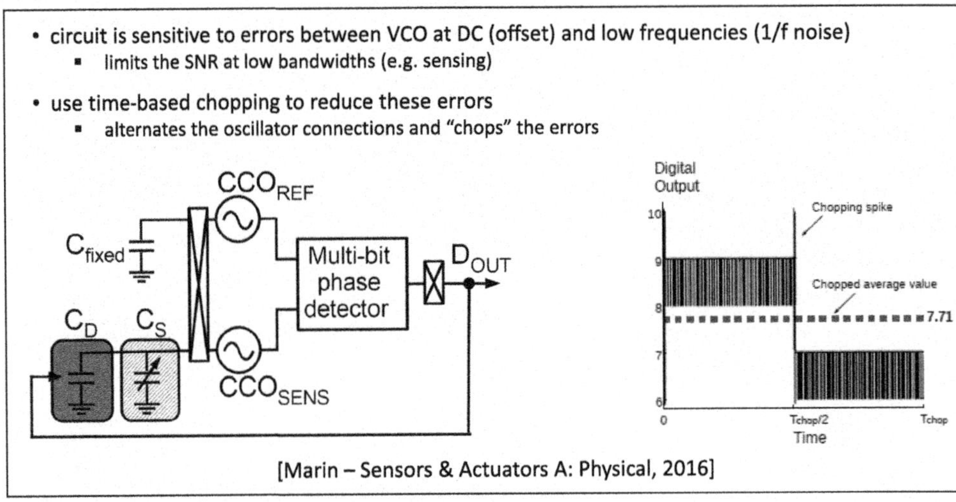

Fig. 6.15 Chopping for low-frequency error reduction (1)

One key limitation is that the architecture relies on the perfect matching of the two oscillators. We take again the basic architecture for the capacitive sensor readout, described in Fig. 6.13. The two capacitance-controlled oscillators (called CCO_{SENS} and CCO_{REF} in the above schematic) are assumed to be perfectly identical. However, we know that they are NOT fully identical in practice, due to mismatches. This causes a stochastic offset between the two oscillators. In addition, since sensing applications typically have a low bandwidth up to a few kHz, this is the region where $1/f$ noise is large. This results in an output spectrum that rapidly increases towards lower frequencies due to the impact of the $1/f$ noise (see the spectrum plots on the next figure Fig. 6.16), hence limiting the effective SNR and resolution achieved in practice.

To address both the offset problem and the $1/f$-noise problem, chopping can be used. As indicated in Fig. 6.15, since this circuit operates in the time domain, this means that the connections to the CCOs are alternated at the rate of the chopping frequency and similarly the output of the phase detector is alternated digitally at the same chopping frequency. This is shown on the right in Fig. 6.15, where chopping results in a chopped average output that is much closer to the correct value than in the non-chopped case. Note also that the time-based chopping does introduce some chopping spikes at the moments of switching, which might create some extra peaks in the output spectrum.

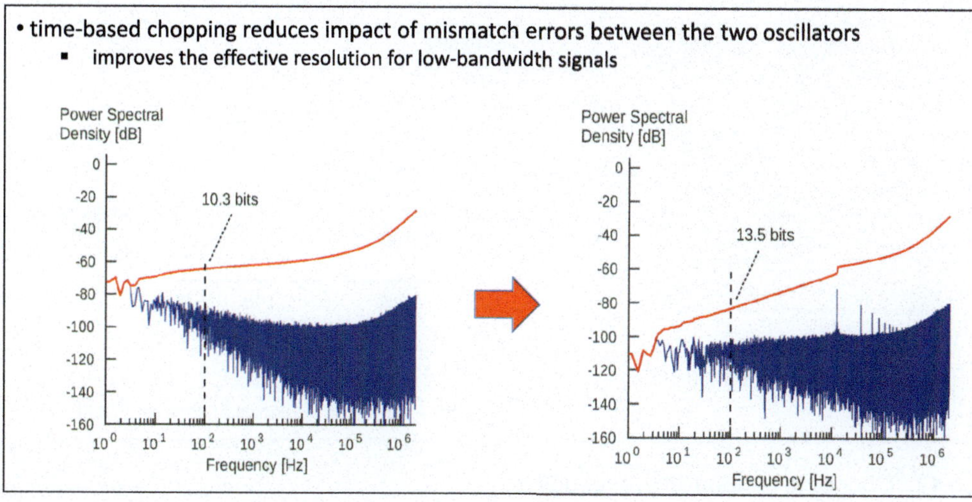

Fig. 6.16 Chopping for low-frequency error reduction (2)

The impact of 1/*f* noise in the oscillators on the VCO-based ADC output spectrum is clearly seen in the spectrum plots in Fig. 6.16. Since sensing applications typically have a low bandwidth up to several kHz, their signals typically fall in the region where 1/*f* noise is large, as clearly shown in the left spectrum plot in Fig. 6.16, hence limiting the effective SNR and resolution achieved in practice (10.3 bits in the example spectrum plot shown).

The time-based chopping introduced in the previous figure Fig. 6.15, largely reduces the impact of the oscillator 1/*f* noise (besides the impact of any oscillator offset), as can be seen in the spectrum plot on the right in Fig. 6.16. The measured spectrum is now a lot flatter at low frequencies, and the SNR with chopping for the same circuit and the same signal bandwidth increases to 13.5 bits in the example shown (compared to only 10.3 bits without chopping).

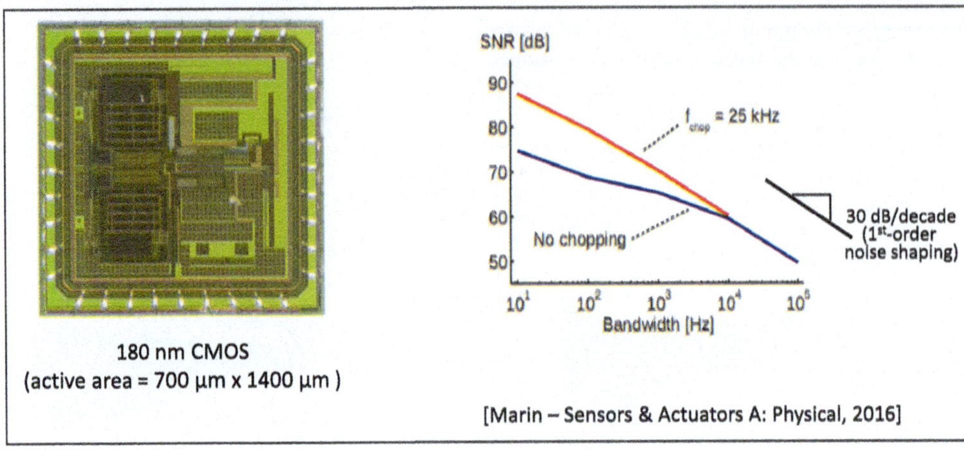

Fig. 6.17 Chopping measurement results

This time-based chopping technique has been implemented and demonstrated experimentally in a prototype chip in 180 nm CMOS technology [Mar16]. The chip photograph is shown on the left in Fig. 6.17. The active area is 0.98 mm^2 (non-optimized). The spectrum plots that were shown in the previous figure are measurement results from this chip with chopping turned off and on.

The figure on the right shows the measured SNR as a function of the signal bandwidth, without and with chopping. For a fixed sampling frequency, the smaller the bandwidth, the larger the oversampling ratio OSR. Since the circuit has first-order noise shaping, we expect that the SNR increases with 30 dB/decade towards higher OSR values (i.e., towards lower bandwidths). However, without the chopping (blue curve), the $1/f$ noise kills the SNR performance and the SNR increases only mildly with decreasing bandwidth (increasing OSR). On the other hand, as shown in the figure, with the chopping turned on (red curve), the impact of the $1/f$ noise reduces, hence restoring the 30 dB/decade increase of the measured SNR with increasing OSR (i.e., with decreasing bandwidth). The chopping frequency used in these measurements is 25 kHz. The results are described in detail in [Mar16].

- first-order and second-order noise-shaping counter-based architecture
- 1.8V supply in 180 nm -- active area = 0.064 mm²
- peak resolution of 16.1 bits

[Sacco - IEEE Journal Solid-State Circuits 2020]

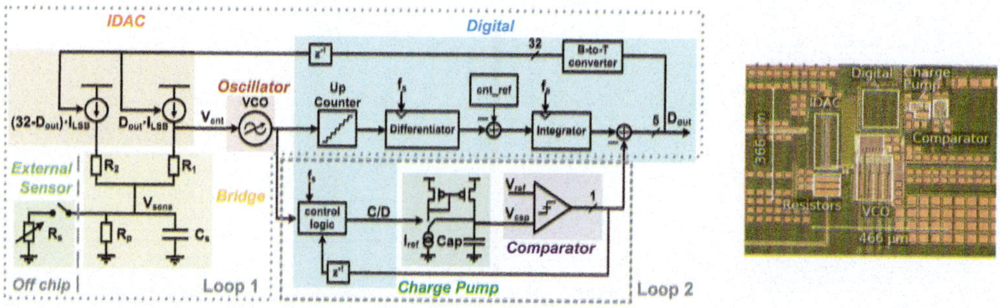

[figures reprinted from IEEE Journal Solid-State Circuits, September 2020 - DOI: 10.1109/JSSC.2020.2987692]

Fig. 6.18 Example: second-order VCO-based resistive sensor readout

Sensor readout architectures with higher orders of noise shaping have also been designed. The circuit in Fig. 6.18 [Sac20] presents a highly digital second-order architecture for a resistive sensor readout that uses only one VCO (instead of two matched VCOs). The single-ended sensor resistor R_S is connected in a bridge configuration (indicated as "Bridge") with the resistors R_p, R_1, and R_2. The two branches R_1 and R_2 are driven by a current-steering DAC (indicated as "IDAC") that itself is driven by the digital output D_{out} of the overall circuit (after binary-to-thermometer conversion). The voltage above resistor R_1 is connected to the VCO input. The VCO transitions are counted to create the actual digital output D_{out}.

In order to create second-order noise shaping, an extra loop (indicated as "Loop 2" on the above schematic) is added. It contains some control logic, a charge pump (where a reference current I_{ref} charges a capacitor Cap in a controlled way) and a comparator. This second loop in 1-1 sturdy MASH (SMASH) configuration creates second-order noise shaping. The circuit is highly digital: all digital parts are indicated in blue on the schematic.

This direct resistive-sensor-to-digital readout circuit has been implemented in 180 nm CMOS with 1.8 V supply. The chip photograph is shown on the right on Fig. 6.18. The active area is only 0.064 mm². Thanks to the second-order noise shaping, the peak resolution has increased to an impressive value of 16.1 bits for 1 ms of conversion time. The chip consumes only 171 μW, resulting in a Walden Figure of Merit (FOM_W) of 2.4 pJ per conversion step, which is excellent for a resistive sensor interface. The specific closed-loop architecture tackles the VCO nonlinearity, achieving more than 14 bits of linearity. Multiple prototype chip samples have been measured in a temperature-controlled environment from −40 °C to 125 °C for the readout of commercial external NTC thermistors: a maximum temperature inaccuracy of 0.3 °C is achieved with only 1-point trimming at room temperature. Since the bridge structure used decouples the sensor excitation from the feedback, also high EMI immunity at the sensor node has been demonstrated. Full details of this circuit architecture can be found in [Sac20].

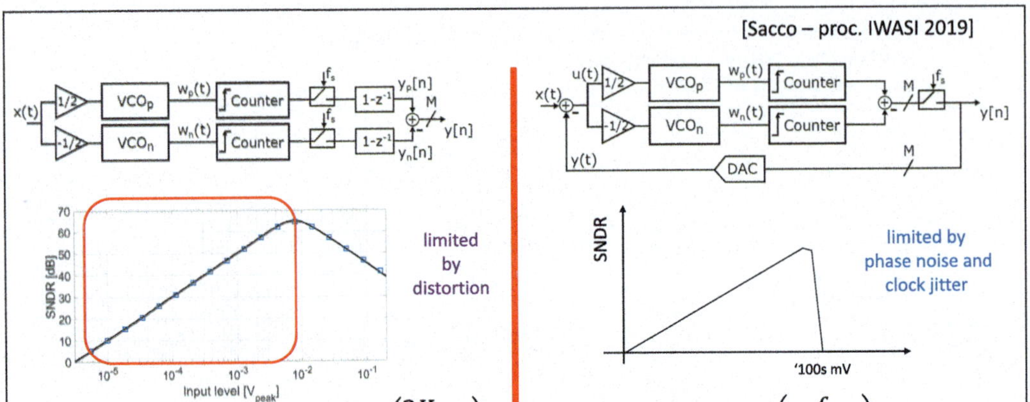

$$SQNR\ [dB] = 6\ log_2\left(\frac{2K_{VCO}}{f_s}\right) - 5.17 + 1.76 + 30\ log_{10}\left(\frac{f_s}{2ABW}\right)$$

Fig. 6.19 Open-loop versus closed-loop architectures

Figure 6.19 summarizes the pros and cons of the two main groups of architectures of VCO-based ADCs: open-loop and closed-loop architectures (as also discussed already in Fig. 3.1). The basic principle schematics are shown at the top. In open loop the input signal typically drives (pseudo-differential) VCOs; the VCO outputs are forwarded to digital decoding blocks (e.g., XOR-flipflop decoder or counter/sampler/differentiator) that sample the VCO signals and the outputs of which are combined to create the final digital output. In closed loop, the sampled output is fed back, converted to analog and subtracted from the inputs. A detailed comparison of both architectures can be found in [Sac19].

The main advantage of open-loop VCO-ADCs is their simplicity as they do not require a sophisticated feedback DAC and most blocks are digital. The main disadvantage is that the nonlinearity of the VCO tuning curve causes distortion that limits the SNDR, as indicated in the SNDR plot in the figure. (The nonlinearity could be improved by either (digital) calibration or some application-specific methods, as discussed in Fig. 6.3).

The main advantage of closed-loop VCO-ADCs is that the high loop gain reduces the impact of the nonlinearity of the VCO tuning curve. As shown on the right in Fig. 6.19, the VCO-ADC's SNDR performance is then limited by the VCO phase noise and the sampling clock jitter sensitivity (e.g., in the DAC). The closed-loop architecture is very power efficient. However, the resulting virtual ground input represents a low-impedance point. In case of a high-impedance source a (power-hungry!) buffer is needed in front of the VCO-ADC.

Reference	Bae20	Shi20	Wu19	Wu20	Xin15	Bab17	Zho20	Mag20	Qui20	Zha20
Year	2020	2020	2019	2020	2015	2017	2020	2020	2020	2020
NTF order	0	0+1	1	1	0+1	**3**	**2**	**1+1**	1	1
analog Correction†	no	FF	no	no	FF	FB	FB	FB	no	FB
Digital calibration	yes	yes	yes	yes	**no**	yes	**no**	**no**	**no**	**no**
Process [nm]	28	16	65	28	40	65	40	65	130	40
Area [mm²]	0.023	0.34	0.244	0.023	0.017	0.01	0.086	0.26	0.04	0.025
Vdd [V]	1.0	1/1.8	1.05	1.2	0.9	1.2	1.1	0.9	1.2/1.5	0.8
BW [MHz]	**2500**	**800**	**200**	40	40	10	5.2	2	0.02	0.01
SNDR [dB]	45.2	58	57	77.8	59.5	65.7	69.4	79.7	73.8	78.5
DR [dB]	50.3	60	60	79.5	62	71	72.3	82.7	97	79
Power [mW]	22.7	280	49.7	10.9	2.57	3.7	0.86	1.25	0.24	0.0047
FoM1 (SNDR) [dB]#	155.6	153	153	**171.4**	161.4	160	167.2	**171.7**	153	**172**
FoM2 (DR) [dB]*	160.7	155	156	**173.1**	163.9	165.3	**170.1**	**174.7**	**176.2**	**172.5**

† FF refers to the input FeedForward technique as explained in section II-B. FB refers to analog FeedBack.
# FoM1 = $SNDR + 10 * log10(BW/Power)$
* FoM2 = $DR + 10 * log10(BW/Power)$

[Gielen, Hernández & Rombouts – IEEE Solid-State Circuits Magazine 2020]

Fig. 6.20 Comparison table with state of the art

Over the past decade, many VCO-based ADCs have been published. This table (Fig. 6.20) from our Solid-State Circuits Magazine paper part 2 [Gie20b] shows a comparison of the reported performances of a representative selection of published VCO-based ADCs, limited to highly digital designs that do not require analog blocks such as opamps. The designs in the table are ordered from very high bandwidth (2.5 GHz) on the left to very low bandwidth (10 kHz) on the right. Apart from their wide variety, an important observation to be made from this table is that all designs are very compact and occupy a very small silicon area.

[Bae20] in the left column is a Nyquist-rate design without noise shaping nor oversampling. With its sampling frequency of 5 GHz it has the highest bandwidth published to date for a VCO-based ADC. A close second is [Shi20] which is a true oversampling converter with an effective application bandwidth of 800 MHz and around 10 effective bits of resolution. Also, the design reported in [Wu20] pushes the bandwidth far in the hundreds of MHz region. It is noteworthy that none of these ultra-high-speed designs uses a global feedback loop, so their noise shaping is only first-order, but they all have digital calibration.

At the other side of the table, there are the extremely efficient designs with audio (or near-audio) bandwidth of [Qui20] and [Sun20b] achieving very good Figure of Merit (FoM) values. In the near-MHz range up to around 40 MHz, several efficient VCO-based ADC designs have been proposed. Here, a much wider variety of design techniques has successfully been deployed: some designs use analog techniques such as closed-loop operation [Zho20] [Mag20] [Zha20] or input feedforwarding [Xin15], whereas others use digital calibration to obtain adequate linearity [Bab17]. Also higher-order noise shaping structures have successfully been demonstrated, like a 1 + 1 MASH structure in [Mag20] or a third-order structure in [Bab17].

Sensor applications are a category on their own: they have a low bandwidth and, when incorporating the sensor directly in the converter, typically also higher power, yet they avoid any buffer or preamplifier. Here are some recent VCO-based examples. [Tu17a], [Tu17b] targets neuron readout, and notwithstanding the low FoM_1 value of 146 dB, it has an excellent sensitivity with a full-scale input of 8 mV_{pp}. [Sac20] presents a highly digital 1-1 sturdy-MASH second-order converter for resistive sensor bridge readout with up to 16.1 bits of resolution and a very good FoM_1 of 163.5 dB.

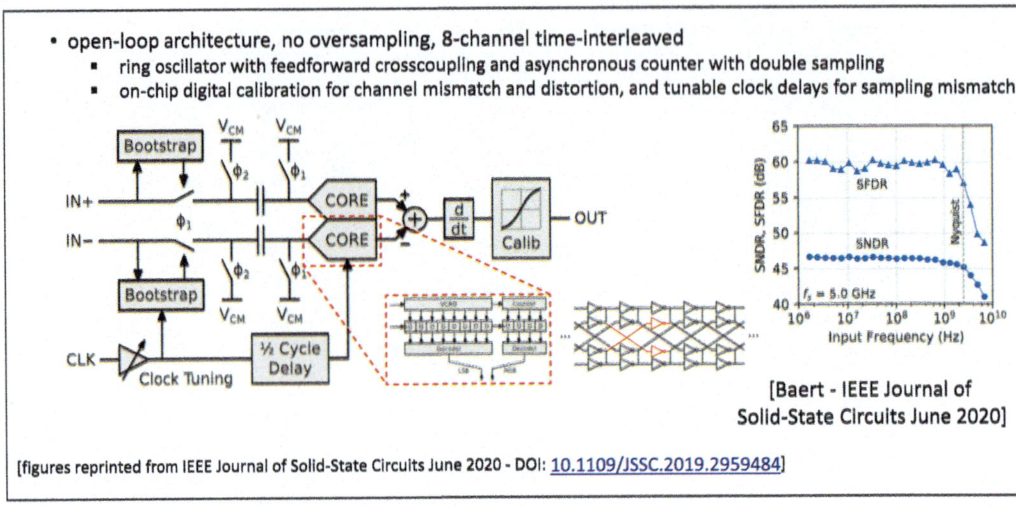

Fig. 6.21 Highest-bandwidth design so far

Figure 6.21 presents the highest-bandwidth VCO-based ADC published so far [Bae20]. It is an open-loop architecture without oversampling. Using eight-channel time-interleaving, it achieves 7.2 effective bits at 5 GS/s in 28 nm CMOS. The circuit schematic is shown on the left. A high-speed ring oscillator with feedforward cross-coupling and a shared tail transistor is combined with an asynchronous counter in order to improve the resolution while minimizing the power consumption. Asynchronous double sampling is used to enable reliable sampling of the asynchronous counter state. On-chip digital calibration is used to compensate for channel mismatch and nonlinear distortion, and sampling time mismatch is corrected using tunable clock delays. With a total power consumption of 22.7 mW, it achieves a Walden Figure of Merit (FOM_W) of 30.5 fJ per conversion step.

The figure on the right shows the measured SNDR and SFDR as a function of the input frequency for a 5-GHz sampling rate. The SNDR peaks at 46.7 dB and drops to 45.2 dB at the Nyquist frequency. The SFDR peaks at 60.3 dB and drops to 57.1 dB at the Nyquist frequency. Beyond the Nyquist frequency, the distortion of the sampling switch becomes the limiting factor. More details about this design can be found in [Bae20].

- hybrid PLL-ΔΣM structure
  - phase locking and phase-frequency detection to perform quantization and dynamic element matching
  - 78.5 dB SNDR for 10 kHz bandwidth →Schreier FoM = 172 dB

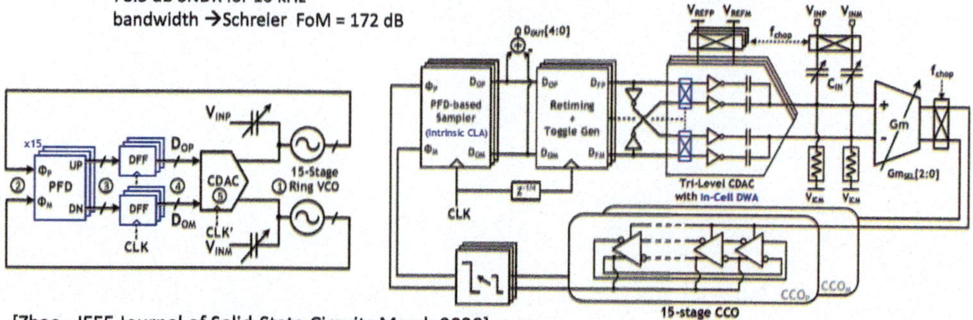

[Zhao - IEEE Journal of Solid-State Circuits March 2020]

[figures reprinted from IEEE Journal of Solid-State Circuits March 2020 - DOI: 10.1109/JSSC.2019.2959479]

Fig. 6.22 Highest-FoM$_S$ design so far

As a last example, Fig. 6.22 presents the VCO-based ADC design with the highest Schreier Figure of Merit (FOM$_S$) published to date [Zha20]. It is a hybrid PLL-ΔΣM structure. The figure on the left shows the block diagram, while the figure on the right shows the more detailed circuit schematic. It is a capacitively coupled VCO-based sensor readout featuring a hybrid phase-locked loop (PLL)-ΔΣ modulator structure. It leverages a phase-locking and phase-frequency detector (PFD) array to concurrently perform quantization and dynamic element matching (DEM), reducing hardware and power compared to counter-based readouts. A low-cost in-cell data-weighted averaging (DWA) scheme is adopted to enable a highly linear tri-level DAC. Fabricated in 40 nm CMOS, the prototype chip achieves 78.5 dB of SNDR in a bandwidth of 10 kHz, consuming 4.68 μW and 0.025 mm^2 of active area. This results in an excellent Schreier Figure of Merit (FOM$_S$) of 172 dB, that advances the state of the art in VCO-based readouts. More details about this design can be found in [Zha20].

Chapter 7
Practical Design Case

G. Gielen et al., *Time-encoding VCO-ADCs for Integrated Systems-on-Chip*,
https://doi.org/10.1007/978-3-030-88067-5_7

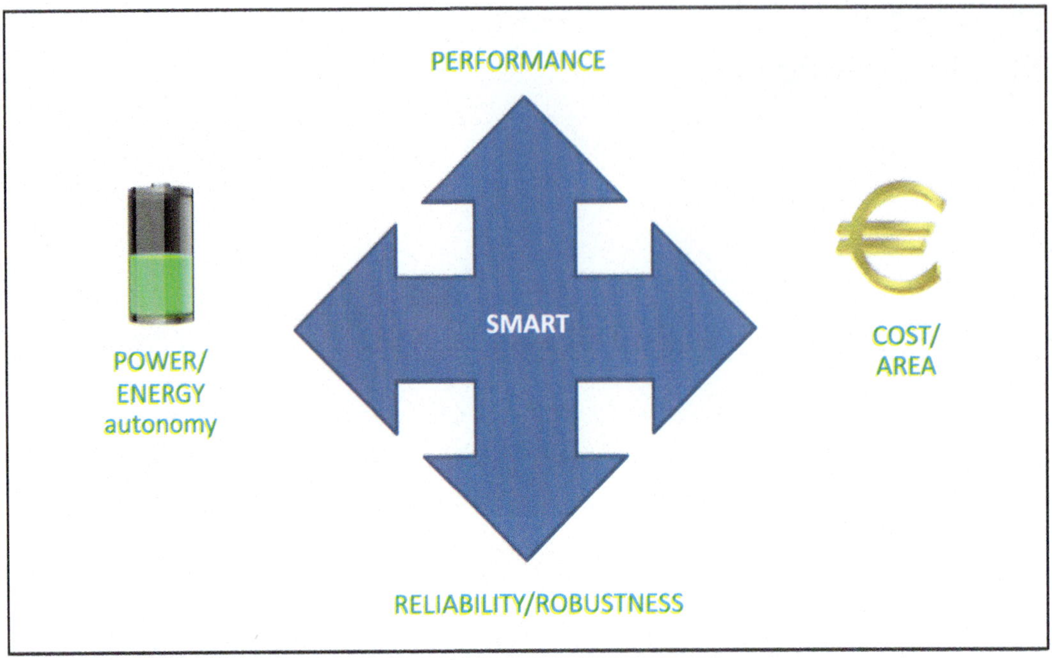

Fig. 7.1 Requirements for sensing circuits

Our world is moving towards a "smart digital society" where an increasing number of applications provide intelligent services with unprecedented functionality to make our daily lives more comfortable in a seamless manner. Many of these applications rely on ubiquitous, distributed sensing to connect the physical world with the electronic world. This happens in almost all application fields, ranging from communications, automotive, and industrial over infrastructure and home control to consumer and healthcare applications. All these applications interact with the real world through sensing and actuation, converting physical signals into electronic signals, and back.

So, what are the requirements for the sensing devices in such applications? (Fig. 7.1)

- First of all, they must meet the technical performance requirements for the application at hand.
- Secondly, they must have a low energy consumption, especially in applications powered from an energy source with limited capacity such as a battery or energy harvesting. This implies that circuits must be power-efficient and only turned on when really needed.
- Thirdly, they must be cheap to make the application at hand affordable. The allowed cost of course differs from, for instance, tagging objects in a warehouse to implanted biomedical devices, to name a few examples. In case of an integrated chip solution for the sensing device, large part of the recurring cost depends on the technology used to fabricate the chip and on the area/size of the chip.
- Fourthly, most sensing applications require a high reliability and robustness against environmental variations (e.g., temperature, supply, humidity, EM interference (EMI), etc.). Safety-critical applications like autonomous vehicles or implanted electronics even require absolute robustness.
- Finally, in recent years, there is also the requirement to make the sensing device smart and adaptive (and highly autonomous where needed).

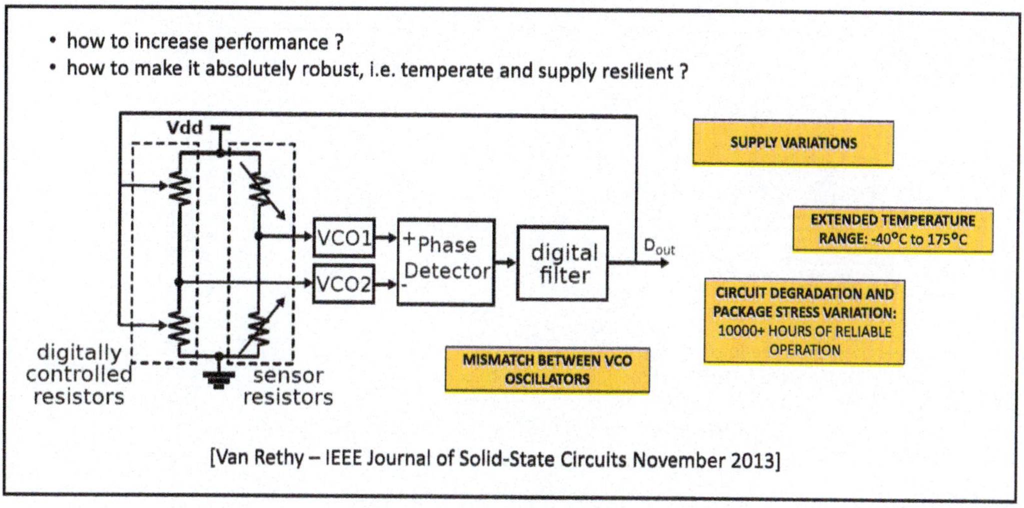

Fig. 7.2 Robust VCO-based resistor-to-digital readout

The amazing achievement in the microelectronics industry over many decades is that the continuous downscaling of the semiconductor technology according to Moore's law helps to achieve the above requirements, both in terms of enabling extra functionality and increasing the level of "intelligence" in many applications, as well as in terms of reducing the cost and physical size.

As an entire practical design case of a VCO-based ADC design for sensing applications, we will describe the design of the robust readout of (differential) resistive sensors. This readout occurs with the VCO-based architecture shown in Fig. 7.2, which has been presented in Fig. 6.12 before and which is described in full detail in [Van13]. The (differential) sensor resistors (right branch) are configured in a Wheatstone-bridge configuration together with digitally controllable (differential) resistors (left branch) that are driven from the digital output D_{out}. The bridge midpoints are forwarded to two twin VCOs, the output phases of which are compared in a phase detector. The high-gain feedback loop drives the phase difference between both VCOs towards zero, hence making both VCOs (on average) run at the same frequency. The digital output needed to achieve this is therefore a digitized version of the sensing signal that influences the value of the sensor resistors.

Using a highly digital VCO-based architecture already results in a small area for this sensor readout circuit. However, how do we make the readout also robust? Robust with respect to any stochastic fabrication tolerances (mismatches in particular), to supply and temperature variations (e.g., the entire range from −40° to 175° in some automotive applications), to aging-related circuit degradation, to package stress variations, and so on. We will show in the next figures how this has been done for this example design case.

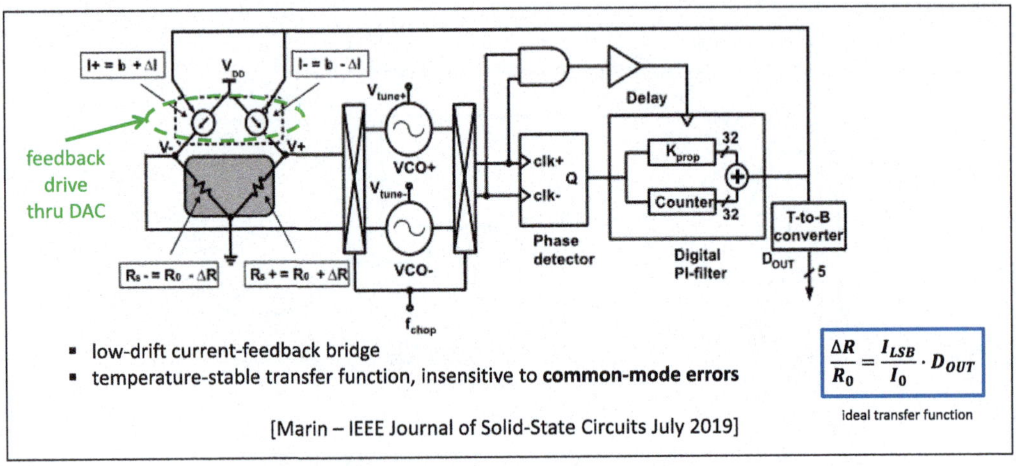

Fig. 7.3 Resilient VCO-based resistive sensor readout (1)

The starting point is to look at the input-output characteristic of the overall circuit and how this is affected by variations like the temperature. To reduce any temperature drift, the principle schematic from Fig. 7.2 is modified as shown above and as described in detail in [Mar19]. The bridge is reconfigured with the differential resistive sensors now at the bottom and the two branches are driven by digitally controlled differential currents. The latter is essentially a small differential current-steering DAC driven by the digital output D_{out}. The remainder of the circuit is essentially the same as in Fig. 7.2. The digital filter is instantiated as a PI-filter (Fig. 7.3).

With this configuration of bridge feedback with a current-steering DAC (IDAC), the ideal overall input-output characteristic of the readout circuit becomes: $\Delta R/R_o = D_{out} \times I_{LSB}/I_o$, where ΔR is the change of the sensing resistors due to the physical signal to be measured, R_o is the nominal value (for zero differential input signal) of the sensing resistors, I_{LSB} is the LSB current of the IDAC, and I_o is the nominal value of the IDAC current for zero digital driving signal D_{out}. The nice aspect of this characteristic is that (in the ideal case) it shows very low drift with temperature, and this across a wide temperature range. The ratio I_{LSB}/I_o is quite stable with temperature, and the configuration with the two (ideally identical) VCOs makes the circuit operation insensitive to common-mode variations like temperature or supply changes that affect both VCOs in the same way.

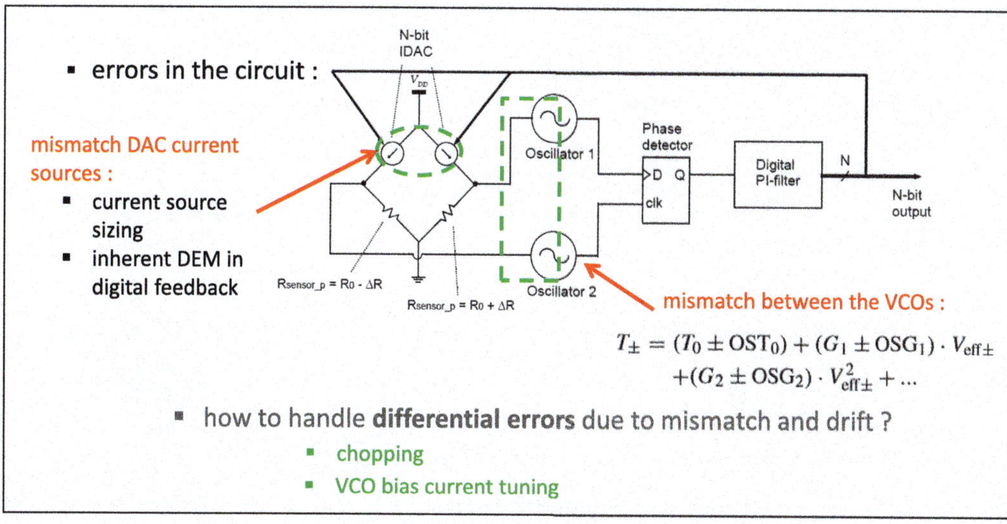

Fig. 7.4 Resilient VCO-based resistive sensor readout (2)

In practical IC implementations, however, manufacturing introduces stochastic mismatches, making the circuit non-ideal. In the considered architecture shown in Fig. 7.4, two mismatches are important.

The first problem is mismatches between the transistors composing the DAC current sources in the IDAC, causing nonlinearity errors. This is addressed by a combination of dynamic element matching that is inherent in the digital output signal D_{out} fed back to the IDAC, and by proper sizing of the current-source transistors for any residual errors.

The second problem is the mismatches between the transistors composing the two VCOs, resulting in mismatch errors between the free-running frequencies f_o of both VCOs, between the conversion gains G_1 of both VCOs, and between any higher-order terms of the tuning characteristics of the two VCOs. This is indicated in the equation shown in Fig. 7.4. These mismatches introduce errors on the overall input-output characteristic of the readout circuit, and also do make the circuit sensitive to common-mode variations like supply and temperature, since the two VCOs no longer track changes identically. To address these VCO errors, two techniques are applied in [Mar19]: chopping and VCO tuning. This will be discussed in detail in the next figures.

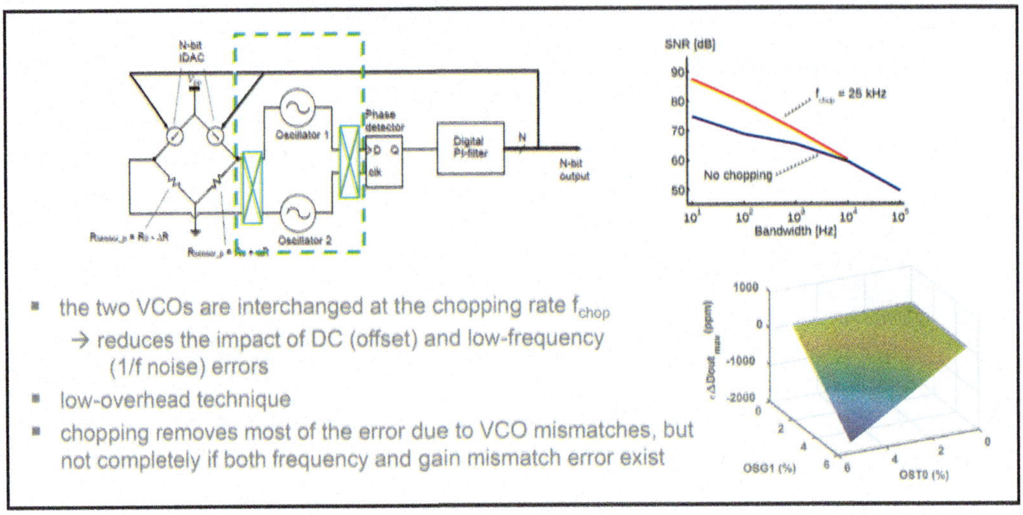

Fig. 7.5 VCO chopping

The first technique applied is chopping (as described in Figs. 6.15 and 6.16 before). At the rate of the chopping frequency f_{chop}, both VCOs are interchanged. To this end chopping circuitry (marked as the green "switching blocks" in the above circuit schematic) is added before and after the VCOs. This continuous interchanging averages out the impact of any offset in the free-running frequency of both VCOs. Note that this is similar to the chopping technique used in standard voltage-mode sensor read-out, be it that this now works in the time domain rather than the voltage domain. A nice feature of this chopping technique is that it has very little hardware overhead.

Since chopping is effective for errors at DC and low frequencies, it also reduces the impact of $1/f$ noise, which can be a major limitation in sensing applications that typically have a bandwidth at low frequencies up to several kHz. As shown in the top right in Fig. 7.5, without chopping (blue curve) the SNR degrades at lower bandwidths due to the $1/f$ noise on top of thermal and quantization noise. Due to the time-based chopping, the impact of $1/f$ noise is reduced and the SNR curve (red curve) is restored to its shape without $1/f$ noise.

Although chopping removes most of the errors due to VCO mismatches, detailed simulations (see bottom right in Fig. 7.5) have shown that the errors are not removed completely when mismatches in both the free-running frequency and the conversion gain exist between the two VCOs. To remove this residual error, another technique needs to be applied, as will be discussed in Fig. 7.6: VCO tuning.

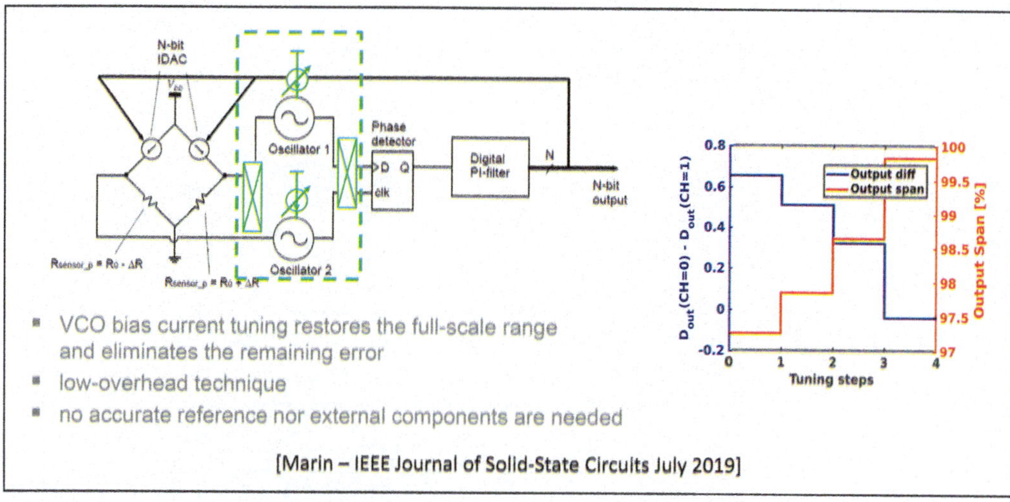

Fig. 7.6 VCO tuning technique

If one of the two error sources (mismatch in free-running oscillation frequency or mismatch in conversion gain) is resolved in a different way, then chopping solves the rest, as discussed in Fig. 7.5. The simple technique proposed in [Mar19] to achieve this is to use on-chip VCO tuning. This is conceptually shown by the tunable current sources in green on the circuit schematic. The VCO implementations are made digitally tunable, such that after tuning both VCOs effectively run at the same frequency. This can be accomplished by applying an arbitrary but fixed input, interchange the two VCOs and compare the output readings: the VCO frequencies are tuned until they show the same output reading for the given fixed input. As shown on the right in Fig. 7.6, when the oscillators are tuned and their outputs become equal (blue curve), the resulting error disappears (red curve).

This simple tuning algorithm can easily be implemented on chip with little overhead. No accurate reference input is needed, nor are any external components needed for this tuning step. When implemented on chip, this procedure could easily be executed again, if needed (e.g., to deal with any differences that would occur due to different aging over time).

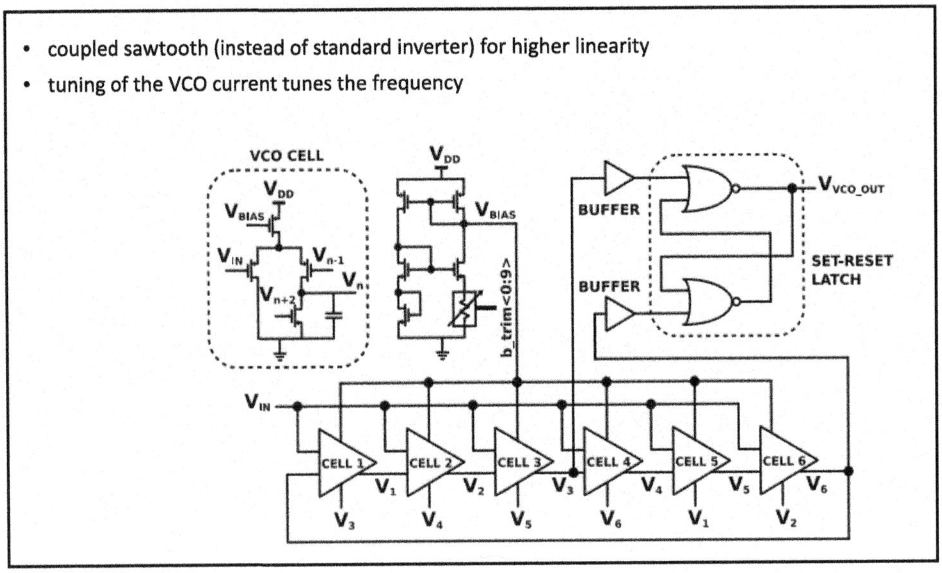

Fig. 7.7 VCO schematic

Figure 7.7 shows the schematic of the VCOs used in the design [Mar19]. The VCO is a standard ring oscillator, but with coupled-sawtooth cells instead of standard inverters for the VCO cells because of their improved linearity. The tuning of the free-running frequency is done by tuning the VCO current for all VCO cells in parallel using the digitally tunable bias voltage V_{BIAS}. The VCO output V_{VCO_OUT} is taken through a set-reset latch with two nodes from the ring oscillator as inputs.

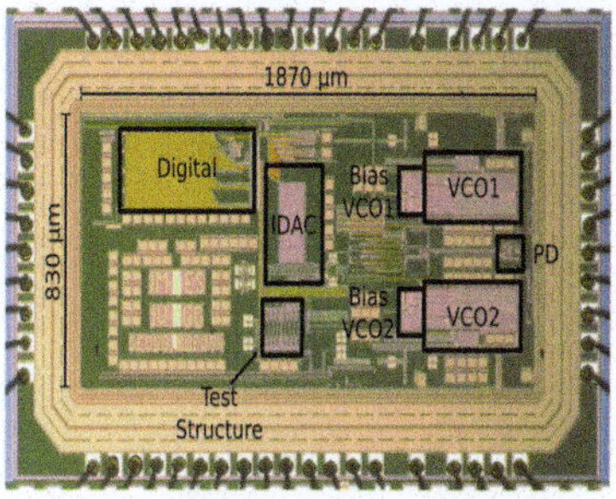

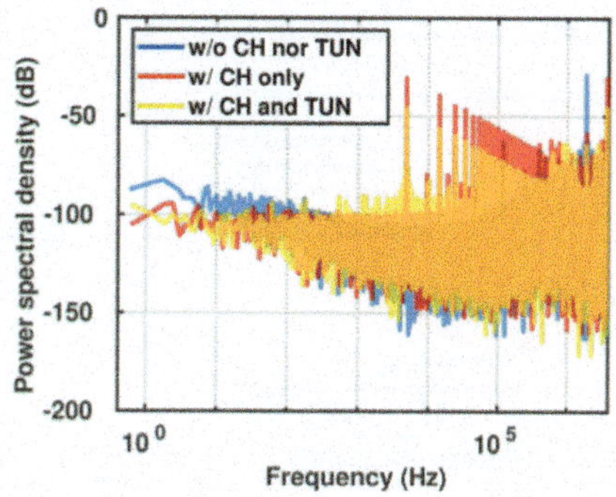

- 15.0 ENOB achieved at 100 ms t_{conv}
- low area of 0.24 mm² in 180nm CMOS
- power of 3.4 mW (for analog 3.3V supply and digital 1.8V supply)
- extremely low drift due to combination of chopping and VCO tuning
  - over temperature range from -40°C to 175°C

[Marin – IEEE Journal of Solid-State Circuits July 2019]

Fig. 7.8 Resistor-to-digital chip implementation

Figure 7.8 shows the resulting chip implementation of the VCO-based resistive sensor readout described in the previous figures. The prototype chip has been implemented in 180 nm CMOS technology. It directly converts the resistor value into a digital output. The chip microphotograph is shown on the left. The supply voltage is 3.3 V for the analog parts and 1.8 V for the digital parts. The chip achieves a measured effective resolution of 15.0 ENOB at 100 ms of conversion time. The active chip area is 0.24 mm² and the power consumption is 3.4 mW. As this was a research prototype, further area and power optimizations are of course possible.

The plot on the right in Fig. 7.8 shows the measured idle-channel output spectrum of the chip in three different settings [Mar19]:

(a) When no chopping nor tuning is applied (blue curve): the impact of the 1/*f* noise at lower frequencies is clearly visible.
(b) When only chopping is used (red curve): the 1/*f* noise is reduced but large errors are seen at higher frequencies.
(c) When both chopping and tuning are used (yellow curve): the 1/*f* noise is reduced and the errors at higher frequencies are reduced too.

The latter setting then results in the measured resolution of 15.0 effective bits.

The chip has been tuned at one temperature (room temperature) and then measured at several other temperatures, showing an extremely low variation across the entire temperature range from −40 °C to +175 °C. These measurement results will be discussed in Fig. 7.9.

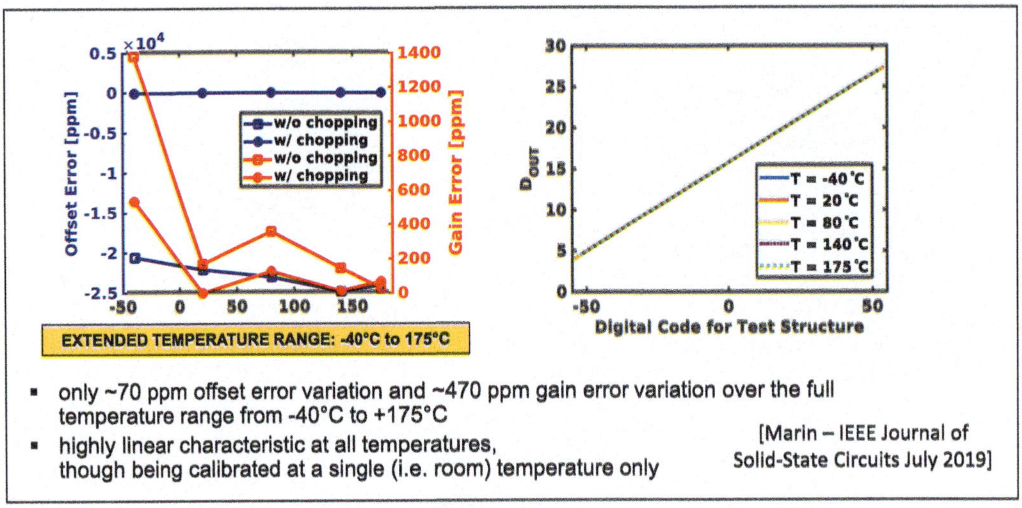

Fig. 7.9 Measured gain and offset variations

Figure 7.9 shows the measurement results of the prototype chip across the entire −40 °C to +175 °C temperature range. The left plot in the figure shows the residual offset and gain errors under the different settings (chopping on and off) after tuning, measured at different temperature values. When chopping is off, the residual errors are large, but the combination of chopping and tuning largely reduces the errors at all temperatures. The residual offset error variation is ~70 ppm across the entire temperature range, while the residual gain error variation is ~470 ppm across the entire temperature range from −40 °C to +175 °C.

The tuning is done at one single temperature (room temperature) and the chip is then used with that setting for the measurements at all other temperatures. The right plot in the figure clearly shows that the circuit response is quite linear and quite insensitive to the temperature across the entire range from −40 °C to +175 °C, even when the chip is tuned at one single temperature only (i.e., at room temperature in practice).

	Mar19	Van13	Wu12	Tu17	Jia17	Jia18	AD17
Architecture (–based)	**Time**	Time	Amplitude	Time	Amplitude	Amplitude	Amplitude
Technology [nm]	**180**	130	700	40	180	180	–
Supply Voltage [V]	**3.3/1.8**	0.85/1.15	5	1.2	1.8	1.8	5
Power [mW]	***3.41**	0.125	1.35	0.02	2.2	2.16	21.5
Active Area [mm²]	****0.26**	0.2	6	0.06	0.73	0.75	–
ENOB	**12.5/15**	8.9	20	12.2	15.4	14.5	24
Temperature Range [°C]	**40 to +175**	-20 to +80	+27 to +50	–	–	–	-40 to +105
Gain Drift [ppm/°C]	**2.3**	56	1.2	–	8.9	–	1
Offset Drift [ppm/°C]***	**0.3**	–	0.25	–	1.25	1.54	0.13
Conversion Time [ms]	**2/100**	0.05	170	0.25	0.5	0.5	2.1

* Power breakdown: $P_{VCOs} = 2.4\text{mW}$; $P_{DAC} = 0.92\text{mW}$; $P_{Digital}$: $91\mu\text{W}$.
** Area breakdown: $A_{VCOs} = 0.075\text{mm}^2$; $A_{DAC} = 0.078\text{mm}^2$; $A_{Digital}$: 0.106mm^2.
*** In the context of time-based conversion the Offset Drift in ppm/°C is a better representative than nV/°C.

20x higher temperature drift resilience (-40°C to 175°C) compared to previous time-based designs & drift resilience wider than voltage-based designs

[Marin – IEEE Journal of Solid-State Circuits July 2019]

Fig. 7.10 Comparison to state of the art

The table in Fig. 7.10 compares the measurement results of the VCO-based resistive sensor readout [Mar19] presented in detail as example design case in the previous figures, with published results from other designs reported in literature. As can be seen in the table, the presented design has a temperature drift resilience across the temperature range from −40 °C to +175 °C that is about 20 times better than previous time-based designs. At the same time, the chip shows a high drift resilience across a wider temperature range than published voltage-based designs.

This shows that VCO-based sensor readout circuits can achieve performance levels, including resilience to ambient variations, that can match or be better than traditional voltage-mode designs, while consuming a low chip area (hence low cost) due to their highly digital nature and their high technology process scalability.

Conclusion

- time-encoding VCO-based ADCs are an interesting ADC architecture in deeply scaled CMOS technologies from sensing (kHz) to high-bandwidth (5 GS/s) applications
  - mostly digital circuits, small area, good power efficiency and high process scalability
- open-loop and closed-loop VCO-based ADCs resemble $\Delta\Sigma$ modulators with first-order noise shaping
  - key system-level design parameters are the rest oscillation frequency and the sampling frequency
  - higher-order solutions are possible by cascading
- to understand subtle system-level effects, Pulse Frequency Modulation theory has to be used
- a twin setup with pseudo-differentially driven VCOs can get good distortion and PSRR performance. Feedback can further improve the linearity and the noise shaping. For open loop and high bandwidths, also feedforward and digital calibration can be used.
- to design the core VCO, a mixture between traditional analog amplifier design and VCO design for PLLs is needed. Several circuit innovations for VCO-based ADCs have been presented in recent years for speed and linearity
- sensor embedding, oversampling, time-based chopping and tuning are among the tricks to improve performance and power efficiency for sensing applications
- performance and resilience levels comparable to or beyond voltage-mode state of the art have been demonstrated

Fig. 1 Take-aways from this book

The above Figure summarizes the major conclusions to take away from this book.

- Time-encoding VCO-based ADCs are an interesting ADC architecture in deeply scaled CMOS technologies for applications ranging from sensing (kHz) to high bandwidths (up to 5 GS/s at this moment). The circuits are mostly digital, have a small silicon area, a good power efficiency, and a high technology process scalability.
- Both open-loop and closed-loop VCO-based ADC architectures are possible. They resemble $\Sigma\Delta$ modulators with implicit first-order noise shaping. The key system-level design parameters are the rest oscillation frequency and the sampling frequency. Higher-order solutions are possible by cascading or extra digital filtering.
- To understand all system effects, Pulse Frequency Modulation theory has to be used to fully grasp the operations occurring on the signals.
- A twin setup with pseudo-differentially driven VCOs can get good distortion and PSRR performance. The use of feedback can further improve the linearity and the noise shaping. For open-loop and high-bandwidth applications, also feedforward and digital calibration can be used as alternatives to reduce the distortion due to the nonlinear VCO tuning characteristic.
- To design the core VCO, a mixture between traditional analog amplifier design and VCO design for PLLs is needed. Several circuit innovations for VCO-based ADCs have been presented in recent years to improve speed and linearity.
- Sensor embedding, oversampling, time-based chopping, and tuning are among the tricks to improve performance and power efficiency for sensing applications.
- Performance and resilience levels comparable or beyond traditional voltage-mode state-of-the-art implementations have been demonstrated, with a lower chip area and good process scalability.

More details can be found in the extensive list of references at the end of this book.

G. Gielen et al., *Time-encoding VCO-ADCs for Integrated Systems-on-Chip*,
https://doi.org/10.1007/978-3-030-88067-5

References

[Abi06] A. A. Abidi. Phase Noise and Jitter in CMOS Ring Oscillators. IEEE J. Solid-State Circuits, 41(8):1803–1816, Aug 2006.

[AD717] AD7193. Analog Devices, Norwood, MA, USA, Feb., 2017.

[Ana16] T. Anand, K. A. A. Makinwa, and P. K. Hanumolu. A VCO Based Highly Digital Temperature Sensor With 0.034 C/mV Supply Sensitivity. IEEE J. Solid-State Circuits, 51(11):2651–2663, Nov 2016.

[Bab15] A. Babaie-Fishani and P. Rombouts. True high-order VCO-based ADC. Electronics Letters, 51(1):23–25, 2015.

[Bab16] A. Babaie-Fishani and P. Rombouts. Highly linear VCO for use in VCO-ADCs. Electronics Letters, 52(4):268–270, 2016.

[Bab17] A. Babaie-Fishani and P. Rombouts. A Mostly Digital VCO-Based CT-SDM With Third-Order Noise Shaping. IEEE Journal of Solid-State Circuits, 52(8):2141–2153, Aug 2017.

[Bae19] M. Baert and W. Dehaene. A 5-GS/s 7.2-ENOB Time-Interleaved VCO-Based ADC Achieving 30.5 fJ/cs. IEEE Journal of Solid-State Circuits, 55(6):1577–1587, 2019

[Bae20] M. Baert and W. Dehaene. A 5-GS/s 7.2-ENOB Time-Interleaved VCO-Based ADC Achieving 30.5 fJ/cs. IEEE J. Solid-State Circuits, 55(6):1577–1587, June 2020.

[Bai95] R. T. Baird and T. S. Fiez. Linearity enhancement of multibit Delta Sigma and D/A converters using data weighted averaging. IEEE Trans. Circuits Syst.-II, 42(12):753–762, Dec. 1995.

[Bay68] E. J. Bayly. Spectral Analysis of Pulse Frequency Modulation in the Nervous Systems. IEEE Transactions on Biomedical Engineering, BME-15(4):257–265, 1968.

[Bor19a] J. Borgmans and P. Rombouts. Enhanced circuit for linear ring VCO-ADCs. Electronics Letters, 55(10):583–585, 2019.

[Bor19b] Jonas Borgmans and Pieter Rombouts. Toward 'digital' analogue-to-digital converters. Electronics Letters, 55(10):568–569, 2019.

[Bor21] Jonas Borgmans, Robbe Riem, and Pieter Rombouts. The Analog Behavior of Pseudo Digital Ring oscillators used in VCO ADCs. IEEE Trans. Circuits Syst.-I: Regular Papers, to be published (accepted for publication), 2021.

[Car18a] Fernando Cardes, Eric Gutierrez, Andres Quintero, Cesare Buffa, Andreas Wiesbauer, and Luis Hernandez. 0.04-mm(2) 103-dB-A Dynamic Range Second-Order VCO-Based Audio Sigma Delta ADC in 0.13-mu m CMOS. IEEE J. Solid-State Circuits, 53(6):1731–1742, Jun. 2018.

[Car18b] Fernando Cardes, Andres Quintero, Eric Gutierrez, Cesare Buffa, Andreas Wiesbauer, and Luis Hernandez. SNDR Limits of Oscillator-Based Sensor Readout Circuits. Sensors, 18(2):445, Feb 2018.

[Car19] F. Cardes, V. Medina, S. Paton, and L. Hernandez. Clock Jitter Analysis of Continuous-Time ΣΔ Modulators Based on a Relative Time-Base Projection. IEEE Transactions on Circuits and Systems I: Regular Papers, 66(3):920–929, 2019.

[Dan10] J. Daniels, W. Dehaene, M. Steyaert, and A. Wiesbauer. A 0.02mm2 65nm CMOS 30MHz BW all-digital differential VCO-based ADC with 64dB SNDR. In 2010 Symposium on VLSI Circuits, pages 155–156, June 2010.

[Dha11] V. Dhanasekaran, M. Gambhir, M. M. Elsayed, E. Sanchez-Sinencio, J. Silva-Martinez, C. Mishra, L. Chen, and E. J. Pankratz. A Continuous Time Multi-BitΔΣADC Using Time Domain Quantizer and Feedback Element. IEEE J. Solid-State Circuits, 46(3):639–650, March 2011.

[Dro12] B. Drost, M. Talegaonkar, and P. K. Hanumolu. Analog Filter Design Using Ring Oscillator Integrators. IEEE J. Solid-State Circuits, 47(12):3120–3129, Dec 2012.

G. Gielen et al., *Time-encoding VCO-ADCs for Integrated Systems-on-Chip*,
https://doi.org/10.1007/978-3-030-88067-5

[Fit47] E. Fitch. The spectrum of modulated pulses. Electrical Engineers - Part IIIA: Radiocommunication, Journal of the Institution of, 94(13):556–564, March 1947.

[Gao12] P. Gao, X. Xing, J. Craninckx, and G. Gielen. Design of an intrinsically-linear double-VCO-based ADC with 2nd-order noise shaping. In 2012 Design, Automation Test in Europe Conference Exhibition (DATE), pages 1215–1220, March 2012.

[Gie16] G. Gielen, J. Van Rethy, J. Marin, M. M. Shulaker, G. Hills, H. P. Wong, and S. Mitra. Time-Based Sensor Interface Circuits in CMOS and Carbon Nanotube Technologies. IEEE Transactions on Circuits and Systems I: Regular Papers, 63(5):577–586, May 2016.

[Gie20a] G. G. E. Gielen, L. Hernandez, and P. Rombouts. Time-Encoding Analog-to-Digital Converters: Bridging the Analog Gap to Advanced Digital CMOS–Part 2: Architectures and Circuits. IEEE Solid-State Circuits Magazine, 12(3):18–27, 2020.

[Gie20b] Georges Gielen, Luis Hernandez, and Pieter Rombouts. Time-encoding analog-to-digital converters: bridging the analog gap to advanced digital CMOS – Part 1: Basic principles. IEEE Solid-State Circuits Magazine, 2020.

[Gut18a] E. Gutierrez, L. Hernandez, F. Cardes, and P. Rombouts. A Pulse Frequency Modulation Interpretation of VCOs Enabling VCOADC Architectures With Extended Noise Shaping. IEEE Transactions on Circuits and Systems I: Regular Papers, 65(2):444–457, Feb 2018.

[Gut18b] E. Gutierrez, P. Rombouts, and L. Hernandez. Why and How VCO-based ADCs can improve instrumentation applications. In 2018 25th IEEE International Conference on Electronics, Circuits and Systems (ICECS), pages 101–104, Dec 2018.

[Her12] L. Hernandez. VCO based multi-stage noise shaping ADC. Electronics Letters, 48(4):206–208, February 2012.

[Her15] L. Hernandez and E. Gutierrez. Analytical Evaluation of VCO-ADC Quantization Noise Spectrum Using Pulse Frequency Modulation. IEEE Signal Processing Letters, 22(2):249–253, Feb 2015.

[Hov97] M. Hovin, A. Olsen, T. S. Lande, and C. Toumazou. Delta-sigma modulators using frequency-modulated intermediate values. IEEE J. Solid-State Circuits, 32(1):13–22, Jan 1997.

[Hua17] S. Huang, N. Egan, D. Kesharwani, F. Opteynde, and M. Ashburn. 28.3 A 125MHz-BW 71.9dB-SNDR VCO-based CT SD ADC with segmented phase-domain ELD compensation in 16nm CMOS. In 2017 IEEE International Solid-State Circuits Conference (ISSCC), pages 470–471, Feb 2017.

[Hua18] J. Huang, F. Laiwalla, J. Lee, L. Cui, V. Leung, A. Nurmikko, and P. P. Mercier. A 0.01-mm2 Mostly Digital Capacitor-Less AFE for Distributed Autonomous Neural Sensor Nodes. IEEE Solid-State Circuits Letters, 1(7):162–165, July 2018.

[Jay19] A. Jayaraj, M. Danesh, S. T. Chandrasekaran, and A. Sanyal. Highly Digital Second-Order $\Delta\Sigma$ VCO ADC. IEEE Trans. Circuits Syst.-I: Regular Papers, 66(7):2415–2425, July 2019.

[Jia17a] Hui Jiang, Kofi A. A. Makinwa, and Stoyan Nihtianov. An energy-efficient 3.7nV/$\sqrt{}$Hz bridge-readout IC with a stable bridge offset compensation scheme. In 2017 IEEE International Solid-State Circuits Conference (ISSCC), pages 172–173, 2017.

[Jia17b] W. Jiang, V. Hokhikyan, H. Chandrakumar, V. Karkare, and D. Markovic. A ±50-mV Linear-Input-Range VCO-Based Neural-Recording Front-End With Digital Nonlinearity Correction. IEEE J. Solid-State Circuits, 52(1):173–184, Jan 2017.

[Jia18] Hui Jiang, Costantino Ligouras, Stoyan Nihtianov, and Kofi A. A. Makinwa. A 4.5 nV/$\sqrt{}$Hz Capacitively Coupled Continuous-Time Sigma-Delta Modulator with an Energy-Efficient Chopping Scheme. IEEE Solid-State Circuits Letters, 1(1):18–21, 2018.

[Kim10] J. Kim, T. K. Jang, Y. G. Yoon, and S. Cho. Analysis and Design of Voltage-Controlled Oscillator Based Analog-to-Digital Converter. IEEE Transactions on Circuits and Systems I: Regular Papers, 57(1):18–30, Jan 2010.

[Kov18] I. Kovacs and M. Neag. New dual-loop topology for ring VCOs based on latched delay cells. In 2018 IEEE International Symposium on Circuits and Systems (ISCAS), pages 1–5, May 2018.

[Laz04] A. A. Lazar and L. T. Toth. Perfect recovery and sensitivity analysis of time encoded bandlimited signals. IEEE Transactions on Circuits and Systems I: Regular Papers, 51(10):2060–2073, Oct 2004.

[Lee15] K. Lee, Y. Yoon, and N. Sun. A Scaling-Friendly Low-Power Small-Area $\Delta\Sigma$ ADC With VCO-Based Integrator and Intrinsic Mismatch Shaping Capability. IEEE Journal on Emerging and Selected Topics in Circuits and Systems, 5(4):561–573, Dec 2015.

[Lee17] L. B. Leene and T. G. Constandinou. Time Domain Processing Techniques Using Ring Oscillator-Based Filter Structures. IEEE Transactions on Circuits and Systems I: Regular Papers, 64(12):3003–3012, Dec 2017.

[Lee19] L. B. Leene and T. G. Constandinou. A 3rd Order Time Domain Delta Sigma Modulator with Extended-Phase Detection. In 2019 IEEE International Symposium on Circuits and Systems (ISCAS), pages 1–5, May 2019.

[Li17] Shaolan Li, Abhishek Mukherjee, and Nan Sun. A 174.3-dB FoM VCO-Based CT Delta Sigma Modulator With a Fully-Digital Phase Extended Quantizer and Tri-Level Resistor DAC in 130-nm CMOS. IEEE J. Solid-State Circuits, 52(7):1940–1952, Jul. 2017.

[Mar16] J. Marin, J. Van Rethy, J. Vergauwen, G. Gielen. Digital-domain chopping technique for high-resolution PLL-based sensor interfaces. Sensors and Actuators A: Physical (Elsevier), 249:294–302, October 2016.

[Mag20] H. Maghami, P. Payandehnia, H. Mirzaie, R. Zanbaghi, H. Zareie, J. Goins, S. Dey, K. Mayaram, and T. S. Fiez. A Highly Linear OTA-Less 1-1 MASH VCO-Based ΔΣ ADC With an Efficient Phase Quantization Noise Extraction Technique. IEEE Journal of Solid-State Circuits, 55(3):706–718, March 2020.

[Mar19] J. Marin, E. Sacco, J. Vergauwen, and G. Gielen. A Robust BBPLL-Based 0.18-μm CMOS Resistive Sensor Interface With High Drift Resilience Over a -40°C–175°C Temperature Range. IEEE J. Solid-State Circuits, 54(7):1862–1873, July 2019.

[Muk19] A. Mukherjee, M. Gandara, B. Xu, S. Li, L. Shen, X. Tang, D. Pan, and N. Sun. A 1-GS/s 20 MHz-BW Capacitive-Input Continuous-Time ΔΣ ADC Using a Novel Parasitic Pole-Mitigated Fully Differential VCO. IEEE Solid-State Circuits Letters, 2(1):1–4, Jan 2019.

[Nai00] R. Naiknaware, Haiming Tang, and T. S. Fiez. Time-referenced single-path multi-bit /spl delta//spl sigma/ adc using a vco-based quantizer. IEEE Transactions on Circuits and Systems II: Analog and Digital Signal Processing, 47(7):596–602, July 2000.

[Nau92] Bram Nauta. A CMOS transconductance-C filter technique for very high frequencies. IEEE J. Solid-State Circuits, 27(2):142–153, 1992.

[Ngu18a] V. Nguyen, F. Schembari, and R. B. Staszewski. A 0.2-V 30-MS/s 11b-ENOB Open-Loop VCO-Based ADC in 28-nm CMOS. IEEE Solid-State Circuits Letters, 1(9):190–193, Sep. 2018.

[Ngu18b] V. Nguyen, F. Schembari, and R. B. Staszewski. A 0.2-V 30-MS/s 11b-ENOB Open-Loop VCO-Based ADC in 28-nm CMOS. 1(9):190–193, Sep. 2018.

[Opt10] F. Opteynde. A maximally-digital radio receiver front-end. In 2010 IEEE International Solid-State Circuits Conference - (ISSCC), pages 450–451, Feb 2010.

[Pay18] P Payandehnia, H Maghami, H Mirzaie, and GC Temes. High-linearity SAR-VCO MASH ΔΣ ADC with second-order noise shaping. Electronics Letters, 54(24):1366–1368, 2018.

[Pep19] F. Pepe and P. Andreani. An Accurate Analysis of Phase Noise in CMOS Ring Oscillators. IEEE Transactions on Circuits and Systems II: Express Briefs, 66(8):1292–1296, Aug 2019.

[Pre11] E. Prefasi, S. Paton, and L. Hernandez. A 7 mW 20 MHz BW Time-Encoding Oversampling Converter Implemented in a 0.08 mm265 nm CMOS Circuit. IEEE J. Solid-State Circuits, 46(7):1562–1574, July 2011.

[Qui18] A. Quintero, C. Perez, E. Gutierrez, and L. Hernandez. VCO-based ADC with a simplified DAC for non-linearity correction. Electronics Letters, 54(13):813–815, 2018.

[Qui20] A. Quintero, C. Buffa, C. Perez, F. Cardes, D. Straeussnigg, A. Wiesbauer, and L. Hernandez. A Coarse-Fine VCO-ADC for MEMS Microphones With Sampling Synchronization by Data Scrambling. IEEE Solid-State Circuits Letters, 3:29–32, 2020.

[Rab97] Shahriar Rabii and Bruce A Wooley. A 1.8-V digital-audio sigma-delta modulator in 0.8-/spl mu/m CMOS. IEEE J. Solid-State Circuits, 32(6):783–796, 1997.

[Rag17] K. Ragab and N. Sun. A 12-b ENOB 2.5-MHz BW VCO-Based 0-1 MASH ADC With Direct Digital Background Calibration. IEEE J. Solid-State Circuits, 52(2):433–447, Feb 2017.

[Rao14] S. Rao, K. Reddy, B. Young, and P. K. Hanumolu. A Deterministic Digital Background Calibration Technique for VCO-Based ADCs. IEEE Journal of Solid-State Circuits, 49(4):950–960, April 2014.

[Raz96] B. Razavi. A study of phase noise in CMOS oscillators. IEEE J. Solid-State Circuits, 31(3):331–343, Mar 1996.

[Raz19] B. Razavi. The Ring Oscillator [A Circuit for All Seasons]. IEEE Solid-State Circuits Magazine, 11(4):10–81, Fall 2019.

[Red12] Karthikeyan Reddy, Sachin Rao, Rajesh Inti, Brian Young, Amr Elshazly, Mrunmay Talegaonkar, and Pavan Kumar Hanumolu. A 16-mW 78-dB SNDR 10-MHz BW CT Delta Sigma ADC Using Residue-Cancelling VCO-Based Quantizer. IEEE J. Solid-State Circuits, 47(12, SI):2916–2927, Dec. 2012.

[Roz97] E. Roza. Analog-to-digital conversion via duty-cycle modulation. IEEE Transactions on Circuits and Systems II: Analog and Digital Signal Processing, 44(11):907–914, Nov 1997.

[Sac19] E. Sacco, J. Vergauwen and G. Gielen. From Open-Loop to Closed-Loop Single-VCO-Based Sensor-to-Digital Converter Architectures: theoretical analysis and comparison. 2019 IEEE 8th International Workshop on Advances in Sensors and Interfaces (IWASI), 2019, pp. 29–34, doi: 10.1109/IWASI.2019.8791296.

[Sac20] E. Sacco, J. Vergauwen, and G. Gielen. A 16.1-bit Resolution 0.064-mm2 Compact Highly Digital Closed-Loop Single-VCO-Based 1-1 Sturdy-MASH Resistance-to-Digital Converter With High Robustness in 180-nm CMOS. IEEE Journal of Solid-State Circuits, page preprint, 2020.

[Sad19] M. Sadollahi and G. C. Temes. A 10-mhz bw 77.9 db sndr dt mash $\delta\sigma$ adc with nc-vco-based quantizer and opamp sharing. IEEE Transactions on Circuits and Systems I: Regular Papers, 66(9):3384–3392, Sep. 2019.

[San17] A. Sanyal and N. Sun. An Energy-Efficient Hybrid SAR-VCO ΔΣ Capacitance-to-Digital Converter in 40-nm CMOS. IEEE J. Solid-State Circuits, 52(7):1966–1976, July 2017.

[Sch05] R. Schreier and G. C. Temes. Understanding Delta-Sigma Data Converters. IEEE, 2005.

[Shi20] H. Shibata, G. Taylor, B. Schell, V. Kozlov, S. Patil, D. Paterson, A. Ganesan, Y. Dong, W. Yang, Y. Yin, Z. Li, P. Shrestha, A. Gopal, A. Bhat, and S. Pavan. 16.6 an 800mhz-bw vco-based continuous-time pipelined adc with inherent anti-aliasing and on-chip digital reconstruction filter. In 2020 IEEE International Solid- State Circuits Conference - (ISSCC), pages 260–262, Feb 2020.

[Str08] M. Z. Straayer and M. H. Perrott. A 12-Bit, 10MHz Bandwidth, Continuous-Time Sigma-Delta ADC With a 5-Bit, 950-MS/s VCO-Based Quantizer. IEEE J. Solid-State Circuits, 43(4):805–814, April 2008.

[Sun18] H. Sun, K. Sobue, K. Hamashita, and U. Moon. An Oversampling Stochastic ADC Using VCO-Based Quantizers. IEEE Transactions on Circuits and Systems I: Regular Papers, 65(12):4037–4050, Dec 2018.

[Taj14] S. Tajima, Y. Okuda, T. Watanabe, H. Aoyama, M. Yamamoto, and T. Uchiyama. High-Resolution Magneto-Impedance Sensor With TAD for Low Noise Signal Processing. IEEE Transactions on Magnetics, 50(11):1–4, Nov 2014.

[Tan18] S. Tannirkulam Chandrasekaran, A. Jayaraj, M. Danesh, and A. Sanyal. A Highly Digital Second-Order Oversampling TDC. IEEE Solid-State Circuits Letters, 1(5):114–117, May 2018.

[Tay10a] G. Taylor and I. Galton. A Mostly-Digital Variable-Rate Continuous-Time Delta-Sigma Modulator ADC. IEEE J. Solid-State Circuits, 45(12):2634–2646, Dec 2010.

[Tay10b] G. Taylor and I. Galton. A Mostly-Digital Variable-Rate Continuous-Time Delta-Sigma Modulator ADC. IEEE J. Solid-State Circuits, 45(12):2634–2646, Dec 2010.

[Tay13] G. Taylor and I. Galton. A Reconfigurable Mostly-Digital Delta-Sigma ADC With a Worst-Case FOM of 160 dB. IEEE Journal of Solid-State Circuits, 48(4):983–995, April 2013.

[Tsi06] Y. Tsividis. Mixed-Domain Systems and Signal Processing Based on Input Decomposition. IEEE Transactions on Circuits and Systems I: Regular Papers, 53(10):2145–2156, Oct 2006.

[Tu17a] C. Tu, Y. Wang, and T. Lin. A Low-Noise Area-Efficient Chopped VCO-Based CTDSM for Sensor Applications in 40-nm CMOS. IEEE J. Solid-State Circuits, 52(10):2523–2532, Oct 2017.

[Tu17b] Chih-Chan Tu, Yu-Kai Wang, and Tsung-Hsien Lin. A 0.06mm2 ±50mV range -82dB THD chopper VCO-based sensor readout circuit in 40nm CMOS. In 2017 Symposium on VLSI Circuits, pages C84–C85, 2017.

[Unn17] V. Unnikrishnan and M. Vesterbacka. Mitigation of Sampling Errors in VCO-Based ADCs. IEEE Transactions on Circuits and Systems I: Regular Papers, 64(7):1730–1739, July 2017.

[Van14] J. Van Rethy, H. Danneels, G. Gielen. Scalable Bang–Bang Phase-Locked-Loop-based integrated sensor interfaces. Elsevier Microelectronics Journal, 45(12):1641–1647, December 2014.

[Van13] Jelle Van Rethy, Hans Danneels, Valentijn De Smedt, Wim Dehaene, and Georges E. Gielen. Supply-Noise-Resilient Design of a BBPLL-Based Force-Balanced Wheatstone Bridge Interface in 130-nm CMOS. IEEE J. Solid-State Circuits, 48(11):2618–2627, NOV 2013.

[Wal99] RH Walden. Analog-to-digital converter survey and analysis. IEEE J. Sel. Areas Commun., 17(4):539–550, Apr. 1999.

[Wat03] T. Watanabe, T. Mizuno, and Y. Makino. An all-digital analog-to-digital converter with 12-μV/LSB using moving-average filtering. IEEE J. Solid-State Circuits, 38(1):120–125, Jan 2003.

[Wat18] T. Watanabe. All-Digital-Very-Scalable-ADC TAD Showing Scaling-Effect in 40/16nm-CMOS Technology. In 2018 25th IEEE International Conference on Electronics, Circuits and Systems (ICECS), pages 105–108, Dec 2018.

[Wu12] Rong Wu, Youngcheol Chae, Johan H. Huijsing, and Kofi A. A. Makinwa. A 20-b ±40-mV Range Read-Out IC With 50-nV Offset and 0.04% Gain Error for Bridge Transducers. IEEE Journal of Solid-State Circuits, 47(9):2152–2163, 2012.

[Wu19] T. Wu and M. S. Chen. A Noise-Shaped VCO-Based Nonuniform Sampling ADC With Phase-Domain Level Crossing. IEEE J. Solid-State Circuits, 54(3):623–635, March 2019.

[Wu20] T. Wu and M. S. Chen. A 40MHz-BW 76.2dB/78.0dB SNDR/DR Noise-Shaping Nonuniform Sampling ADC with Single Phase-Domain Level Crossing and Embedded Nonuniform Digital Signal Processor in 28nm CMOS. In 2020 IEEE International Solid-State Circuits Conference - (ISSCC), pages 262–264, Feb 2020.

[Xie19] Y. Xie, Y. Liang, M. Liu, S. Liu, and Z. Zhu. A 10-Bit 5 MS/s VCO-SAR ADC in 0.18-μm CMOS. IEEE Trans. Circuits Syst.-II: Express Briefs, 66(1):26–30, Jan 2019.

[Xin13] X. Xing, P. Gao and G. Gielen. A 40MHz-BW two-step open-loop VCO-based ADC with 42fJ/step FoM in 40nm CMOS. 2013 Proceedings of the ESSCIRC (ESSCIRC), 2013, pp. 327–330, doi: 10.1109/ESSCIRC.2013.6649139.

[Xin15] X. Xing and G. G. E. Gielen. A 42 fJ/Step-FoM Two-Step VCO-Based Delta-Sigma ADC in 40 nm CMOS. IEEE J. Solid-State Circuits, 50(3):714–723, March 2015.

[Xu19] H. Xu and A. A. Abidi. Analysis and Design of Regenerative Comparators for Low Offset and Noise. IEEE Transactions on Circuits and Systems I: Regular Papers, 66(8):2817–2830, Aug 2019.

[Yeo18] P. Yeon, M. S. Bakir, and M. Ghovanloo. Towards a 1.1 mm2 free-floating wireless implantable neural recording SoC. In 2018 IEEE Custom Integrated Circuits Conference (CICC), pages 1–4, April 2018.

[Yu13] W. Yu, J. Kim, K. Kim, and S. Cho. A Time-Domain High-Order MASH ΔΣ ADC Using Voltage-Controlled Gated-Ring Oscillator. IEEE Transactions on Circuits and Systems I: Regular Papers, 60(4):856–866, 2013.

[Zha20] W. Zhao, S. Li, B. Xu, X. Yang, X. Tang, L. Shen, N. Lu, D. Z. Pan, and N. Sun. A 0.025-mm2 0.8-V 78.5-dB SNDR VCO-Based Sensor Readout Circuit in a Hybrid PLL-ΔΣ M Structure. IEEE Journal of Solid-State Circuits, 55(3):666–679, March 2020.

[Zho20] Y. Zhong, S. Li, X. Tang, L. Shen, W. Zhao, S. Wu, and N. Sun. A Second-Order Purely VCO-Based CT ΔΣ ADC Using a Modified DPLL Structure in 40-nm CMOS. IEEE Journal of Solid-State Circuits, 55(2):356–368, Feb 2020.

Index

G. Gielen et al., *Time-encoding VCO-ADCs for Integrated Systems-on-Chip*,
https://doi.org/10.1007/978-3-030-88067-5

The manufacturer's authorised representative in the EU is Springer
Nature Customer Service Centre GmbH, Europaplatz 3, 69115 Heidelberg,
Germany. If you have any concerns regarding our products, please
contact ProductSafety@springernature.com

Printed and bound by CPI Group (UK) Ltd, Croydon, CR0 4YY
30/07/2026
02182086-0005